Numerical Modeling in Science and Engineering

Numerical Modeling in Science and Engineering

Editor: Lawrence Bech

New York

Published by NY Research Press
118-35 Queens Blvd., Suite 400,
Forest Hills, NY 11375, USA
www.nyresearchpress.com

Numerical Modeling in Science and Engineering
Edited by Lawrence Bech

International Standard Book Number: 978-1-63238-542-0 (Hardback)

Cataloging-in-Publication Data

Numerical modeling in science and engineering / edited by Lawrence Bech.
    p. cm.
Includes bibliographical references and index.
ISBN 978-1-63238-542-0
1. Mathematical models. 2. Science--Mathematical models.
3. Engineering--Mathematical models. 4. Numerical analysis. I. Bech, Lawrence.
TA342 .N86 2017
511.8--dc23

Printed in the United States of America.

Contents

Preface.. VII

Chapter 1 **Mathematical modeling and dynamic simulation of a class of drive systems with permanent magnet synchronous motors**........................ 1
M. Mikhova

Chapter 2 **Finite element for non-stationary problems of viscoelastic orthotropic beams**.. 9
M. Zajíček, V. Adámek and J. Dupal

Chapter 3 **An evaluation of the stress intensity factor in functionally graded materials**.. 21
M. Ševčík, P. Hutař, L. Náhlík and Z. Knésl

Chapter 4 **Mathematical modeling of a biogenous filter cake and identification of oilseed material parameters**...................................... 31
J. Očenášek and J. Voldřich

Chapter 5 **Analysis of composite car bumper reinforcement**........................... 43
V. Kleisner and R. Zemčík

Chapter 6 **A refined description of the crack tip stress field in wedge-splitting specimens – a two-parameter fracture mechanics approach**.............. 53
S. Seitl, Z. Knésl, V. Veselý and L. Řoutil

Chapter 7 **Influence of crucial parameters of the system of an inverted pendulum driven by fibres on its dynamic behaviour**.................... 69
P. Polach, M. Hajžman and Z. Šika

Chapter 8 **Linearization of friction effects in vibration of two rotating blades**......... 81
M. Byrtus, M. Hajžman and V. Zeman

Chapter 9 **Various methods of numerical estimation of generalized stress intensity factors of bi-material notches**.. 99
J. Klusák, T. Profant and M. Kotoul

Chapter 10 **Reconstruction of a fracture process zone during tensile failure of quasi-brittle materials**... 107
V. Veselý and P. Frantík

Chapter 11 **Utilization of random process spectral properties for the calculation of fatigue life under combined loading**...................................... 121
J. Svoboda, M. Balda and V. Fröhlich

Chapter 12   **Numerical simulation of plastic deformation of aluminium workpiece induced by ECAP technology**...131
R. Melichera

Chapter 13   **Analytical and numerical investigation of trolleybus vertical dynamics on an artificial test track**...143
P. Polach, M. Hajžman, J. Soukup and J. Volek

Chapter 14   **Dynamic wheelset drive load of the railway vehicle caused by short-circuit motor moment**...155
V. Zeman and Z. Hlaváča

Chapter 15   **Implementation of skeletal muscle model with advanced activation control**... 167
H. Kocková and R. Cimrman

Chapter 16   **Numerical approximation of flow in a symmetric channel with vibrating walls**... 181
P. Sváček and J. Horáček

Chapter 17   **Validation of the adjusted strength criterion LaRC04 for uni-directional composite under combination of tension and pressure**..193
J. Krystek, R. Kottner, L. Bek and V. Laš

Permissions

List of Contributors

Index

Preface

This book aims to highlight the current researches and provides a platform to further the scope of innovations in this area. This book is a product of the combined efforts of many researchers and scientists from different parts of the world. The objective of this book is to provide the readers with the latest information in the field.

Numerical modeling is the use of computational and mathematical methods to solve scientific problems. This book on numerical modeling in science and engineering deals with the basic elements that are dealt by this field such as algorithmic simulation, computer software development and computational intelligence. There has been rapid progress in this field and its applications are finding their way across multiple industries. This book studies, analyzes and upholds the pillars of numerical modeling and its utmost significance in modern times. While understanding the long-term perspective of the topics, the book makes an effort in highlighting their impact as a modern tool for the growth of the discipline. It will help the readers in keeping pace with the rapid changes in this field.

I would like to express my sincere thanks to the authors for their dedicated efforts in the completion of this book. I acknowledge the efforts of the publisher for providing constant support. Lastly, I would like to thank my family for their support in all academic endeavors.

Editor

Mathematical Modeling and Dynamic Simulation of a Class of Drive Systems with Permanent Magnet Synchronous Motors

M. Mikhov[a,*]

[a] *Faculty of Automatics, Technical University of Sofia, 8 Kliment Ohridski Blvd., 1757 Sofia, Bulgaria*

Abstract

The performance of a two-coordinate drive system with permanent magnet synchronous motors is analyzed and discussed in this paper. Both motors have been controlled in brushless DC motor mode in accordance with the rotor positions. Detailed study has been carried out by means of mathematical modeling and computer simulation for the respective transient and steady-state regimes at various load and work conditions. The research carried out as well as the results obtained can be used in the design, optimization and tuning of such types of drive systems. They could be also applied in the teaching process.

Keywords: two-coordinate drive, brushless DC motor control mode, dynamic simulation

1. Introduction

A number of applications in the mechanical industry require two-coordinate drive systems with good static, dynamic and energetic characteristics. In recent years, permanent magnet synchronous motors (PMSM) have been used in such cases, because they have some advantages, including: compact form; low power loss and high efficiency; high power/mass ratio; good heat dissipation characteristics; low rotor inertia and good dynamics; high speed capabilities.

Applying some control methods, drive systems with such motors can combine the advantages of both DC and AC motor systems. For example, the PMSM electric drives performance improves significantly if their control is executed according to the rotor position. Thus commutation flexibility is provided, avoiding the risk of synchronization failure in case of overload [1, 2, 7, 8].

The complexity of electromechanical systems makes them difficult for description and study in some cases. For this reason, mathematical modeling and computer simulation are widely applied for the purpose of their analysis and synthesis. Such an approach provides for very good conditions to investigate electric drives behavior in various transient and steady-state working regimes, which is not always convenient or possible in industrial and laboratory environments [3, 4, 5, 6].

This paper considers a two-coordinate electromechanical system with permanent magnet synchronous motors, which are controlled in brushless DC motor mode. Detailed study has been carried out by means of modeling and computer simulation for the respective dynamic and static regimes at different operation modes. Behavior analysis has been made aiming at improvement of the drive system performance.

*Corresponding author. e-mail: mikhov@tu-sofia.bg.

2. Mathematical modeling of the drive system

2.1. Features of the drive system

A simplified block diagram of the drive system under consideration is represented in fig. 1, where the notations are as follows: PC – position controller; SC1, SC2 – speed controllers; RC1, RC2 – current reference blocks; CC1, CC2 – three-phase current controllers; IC1, IC2 – inverter control blocks; VI1, VI2 – voltage source inverters; UR – uncontrollable rectifier; C – filter capacitor; PS1, PS2 – position sensors; PF1, PF2 – position feedback blocks; SF1, SF2 – speed feedback blocks; M1, M2 – motors; L1, L2 – loads at the respective coordinate axes; V_{sr_1}, V_{sr_2} – speed reference signals; V_{cr_1}, V_{cr_2} – current reference signals; V_{pf_1}, V_{pf_2} – position feedback signals; V_{sf_1}, V_{sf_2} – speed feedback signals; V_{cf_1}, V_{cf_2} – current feedback signals; V_d – DC link voltage; ω_1, ω_2 – motor speeds; θ_1, θ_2 – angular positions; T_{l_1}, T_{l_2} – load torques applied to the respective motor shafts.

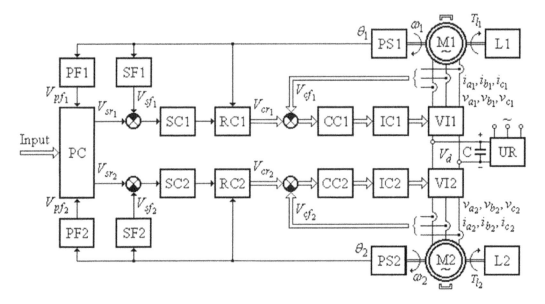

Fig. 1. Block diagram of the drive system under consideration

The motors used are characterized by having permanent magnet-produced fields on the rotors and armature windings on the stators. Both subsystems have identical cascade structures with subordinate regulation. The motors have been controlled in brushless DC motor mode in accordance with the respective rotor positions. Control loops optimization and tuning have been done sequentially, starting from the innermost ones.

2.2. Mathematical description

In order to obtain a suitable simulation model some assumptions have been made, such as:

- motors are unsaturated;

- eddy-currents and hysteresis effects have negligible influence on the stator currents;

- motors are symmetrical three-phase machines;

- self and mutual inductances are constant and independent of the rotor position;

- devices in the power-electronic circuits are ideal.

Fig. 2 shows the equivalent circuit diagram of PMSM and the voltage source inverter for one coordinate axis. The corresponding notations are as follows: S1÷S6 – power electronic switches; D1÷D6 – freewheeling diodes; v_a, v_b, v_c – voltages applied on stator phases a, b and c respectively; i_a, i_b, i_c – phase currents; e_a, e_b, e_c – back EMF voltages; R_a, R_b, R_c – stator phase resistances; L_a, L_b, L_c – phase self inductances; L_{ba}, L_{ca}, L_{cb} – mutual inductances.

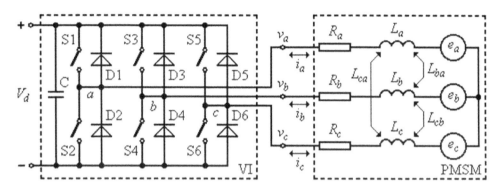

Fig. 2. Equivalent circuit diagram of PMSM and the voltage source inverter

The vector-matrix model of such a set is described as follows:

$$
\begin{bmatrix} v_a \\ v_b \\ v_c \end{bmatrix} = \begin{bmatrix} R_s & 0 & 0 \\ 0 & R_s & 0 \\ 0 & 0 & R_s \end{bmatrix} \begin{bmatrix} i_a \\ i_b \\ i_c \end{bmatrix} + \begin{bmatrix} L_a & L_{ba} & L_{ca} \\ L_{ba} & L_b & L_{cb} \\ L_{ca} & L_{cb} & L_c \end{bmatrix} \frac{\mathrm{d}}{\mathrm{d}t} \begin{bmatrix} i_a \\ i_b \\ i_c \end{bmatrix} + \begin{bmatrix} e_a \\ e_b \\ e_c \end{bmatrix} , \quad (1)
$$

where $R_s = R_a = R_b = R_c$ is the stator phase resistance.

The back EMF voltage waveforms are expressed by the equation:

$$
\begin{bmatrix} e_a \\ e_b \\ e_c \end{bmatrix} = \omega \frac{\mathrm{d}}{\mathrm{d}\theta} \begin{bmatrix} \Phi_a \\ \Phi_b \\ \Phi_c \end{bmatrix} , \quad (2)
$$

where Φ_a, Φ_b, Φ_c are the stator magnetic fluxes of the motor phases a, b and c respectively.

Since the motor windings are star connected (fig. 2), the next relation is valid:

$$
i_a + i_b + i_c = 0. \quad (3)
$$

Assuming there is no change in the rotor reluctance with angular position, then:

$$
\begin{aligned}
L_a &= L_b = L_c = L; \quad &(4) \\
L_{ba} &= L_{ca} = L_{cb} = M. \quad &(5)
\end{aligned}
$$

Taking into consideration Eqs. (4) and (5), the vector-matrix Eq. (1) is arranged as follows:

$$
\begin{bmatrix} v_a \\ v_b \\ v_c \end{bmatrix} = R_s \begin{bmatrix} i_a \\ i_b \\ i_c \end{bmatrix} + L_s \frac{\mathrm{d}}{\mathrm{d}t} \begin{bmatrix} i_a \\ i_b \\ i_c \end{bmatrix} + \begin{bmatrix} e_a \\ e_b \\ e_c \end{bmatrix} , \quad (6)
$$

where $L_s = L - M$ is the stator phase inductance.

The motor electromagnetic torque can be expressed as:

$$T = \frac{e_a i_a + e_b i_b + e_c i_c}{\omega}, \tag{7}$$

and the mechanical dynamics equations are as follows:

$$J\frac{d\omega}{dt} = T - T_l - D\omega; \tag{8}$$

$$\frac{d\theta}{dt} = \omega, \tag{9}$$

where J is the total inertia referred to the respective motor shaft; D – the viscous damping coefficient.

3. Computer simulation and performance analysis

Every specific working regime of the two-coordinate drive system requires an appropriate control algorithm. For this reason, using the MATLAB/SIMULINK software package some computer simulation models have been developed to analyze the system under consideration.

Detailed investigation has been carried out by means of mathematical modeling and computer simulation for the respective transient and steady-state regimes at various working conditions.

3.1. Phase currents formation

The current controllers have a programmable hysteresis band which determines the respective modulation frequency.

Fig. 3 illustrates the principle of current pulses formation. The used notations are as follows: i_r – reference phase current waveform; $2\Delta i_r = i_{max} - i_{min}$ – reference hysteresis band; i_{max} and i_{min} – maximum and minimum current values, respectively; $T_m = t_{on} + t_{off}$ – modulation period.

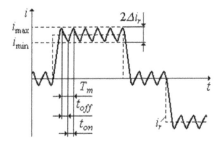

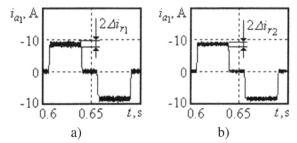

a)   b)

Fig. 3. Principle of current pulses formation   Fig. 4. Influence of the hysteresis band

The modulation frequency is expressed by:

$$f_m = \frac{1}{T_m}. \tag{10}$$

The hysteresis band influence of the first current controller is shown in fig. 4. It represents the phase current waveforms for two different hysteresis bands ($2\Delta i_{r_1}$ and $2\Delta i_{r_2}$ respectively). As evident, narrowing the zone, i.e. increasing the chopping frequency will result in reduction of the current pulsations.

The analysis carried out shows that this frequency is not of a constant value and depends on the following factors:

- hysteresis band;
- reference motor speed;
- power circuit electromagnetic time-constant;
- initial current value.

This analysis allows selecting of the optimal hysteresis band in accordance with the respective power switches. In the research carried out the maximum modulation frequency has been limited to a value of $f_{m\,max} < 8\,000$ Hz.

Fig. 5 illustrates the dynamic maintenance of zero phase current in the respective interval by means of pulse width modulation.

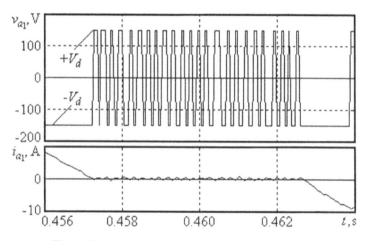

Fig. 5. Dynamic maintenance of zero phase current

Fig. 6 shows the phase current waveform i_{a_1} and the respective trapezoidal back EMF voltage e_{a_1}. In this case the simulation results have been obtained for a quasi-steady-state regime at rated load applied to the motor shaft.

The three-phase current waveforms i_{a_1}, i_{b_1}, and i_{c_1} for the same load torque are represented in fig. 7.

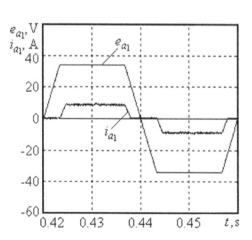

Fig. 6. Phase current and back EMF voltage

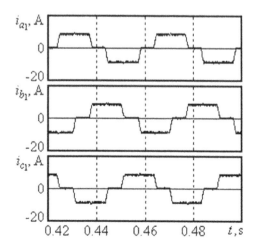

Fig. 7. Three-phase current waveforms

3.2. Breaking regime

Various breaking modes have been investigated aiming at fast and efficient stopping and reversing. Some time-diagrams obtained at reverse speed control with electrical braking are shown in fig. 8, where T_{b1} is the braking torque. The reference speed values in this case are $\pm\omega_{1r} = \pm157$ rad/s. As evident, the braking mode applied ensures good dynamics with maximum reverse time less than 0.1 s.

3.3. Current limitation and speed stabilization

The phase currents limitation is provided through the current reference signals, formed by the respective speed controllers. These controllers have been optimized in such a way, that the static errors possibly caused by disturbances are eliminated.

Fig. 9 illustrates motor speed stabilization when the load torque changes. The respective transient and steady-state regimes are represented, as well as the drive system reaction to a disturbance expressed as load change.

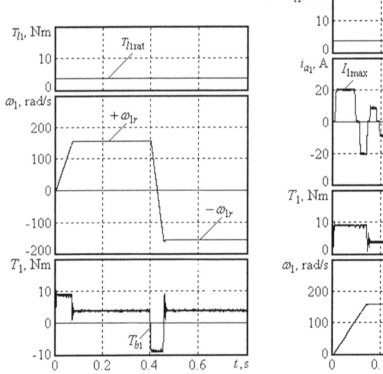

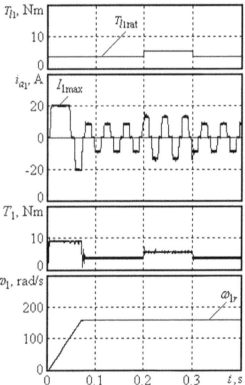

Fig. 8. Reversing with electrical breaking   Fig. 9. Current limitation and speed stabilization

In this case the load torque applied is equal to the rated value $T_{l1\mathrm{rat}}$ and the respective disturbance is $\Delta T_{l_1} = 0.5T_{l_1\mathrm{rat}}$. The load change is applied during the time interval of $t = (0.2\ \mathrm{s} \div 0.3\ \mathrm{s})$. As evident, at an appropriate tuning of the speed controller, static speed error does not appear when T_{l_1} changes. The reference speed is $\omega_{1r} = 157$ rad/s and the starting current is limited to the maximum admissible value of $I_{1\max}$, which provides for a maximum starting motor torque.

3.4. Two-coordinate position control

Two algorithms for trajectory formation have been investigated, namely simultaneous and consecutive movements along both coordinate axes.

Fig. 10 represents some simulation results obtained for two-coordinate position control with consecutive motion along the coordinate axes. The reference distances are indicated with θ_{1r} and θ_{2r} respectively.

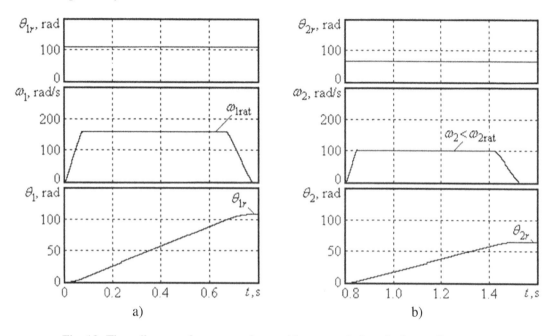

Fig. 10. Time-diagrams for consecutive position control along both coordinate axes

Fig. 10a illustrates the realization of a reference motion along the first coordinate axis. The respective motor works at maximum speed of $\omega_1 = \omega_{1\text{rat}}$, which ensures fast operation of the drive system.

Fig. 10b shows the subsequent positioning along the second coordinate axis. In this case the movement is carried out with a lower speed of $\omega_2 < \omega_{2\text{rat}}$.

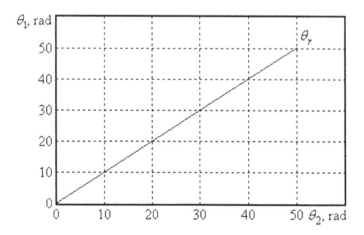

Fig. 11. Trajectory of simultaneous movement along both coordinate axes

A simultaneous movement along both coordinate axes is illustrated in fig. 11, when the two motor speeds are equal.

The motors used are identical, both with rated power $P_{rat} = 0.6$ kW and rated angular speed $\omega_{rat} = 157$ rad/s.

4. Conclusion

The performance of a two-coordinate drive system with permanent magnet synchronous motors has been discussed in this paper. Both motors have been controlled in brushless DC motor mode in accordance with the respective rotor positions.

Using the MATLAB/SIMULINK software package a number of computer simulation models have been developed to analyze this class of drive systems. Detailed study has been carried out for the dynamic and static regimes at various loads, disturbances and work conditions.

The analysis shows that the control approach applied ensures good performance, which makes it suitable for a variety of applications in the mechanical industry.

The research carried out as well as the results obtained can be used in the design, optimization and tuning of such types of drive systems. They could be also applied in the teaching process.

Acknowledgements

This work has been supported by the Technical University of Sofia under the research project No. 091NI142-08/2009.

References

[1] Boldea, I., Nasar, S. A., Electric drives, CRC, Boca Raton, 1999.
[2] Chen, C., Cheng, M., A new cost effective sensorless commutation method for brushless DC motors without phase shift circuit and neutral voltage, IEEE Trans. on Power Electronics, 2 (22) (2007) 644–653.
[3] Mikhov, M. R., Investigation of a permanent magnet synchronous motor control system, Technical Ideas, 3/4 (38) (2001) 23–34.
[4] Mikhov, M. R., Analysis and simulation of a servo drive system with hysteresis current control, Proceedings of the International Conference on Automatics and Informatics, Sofia, Bulgaria, 2002, pp. 197–200.
[5] Ong, C., Dynamic simulation of electric machinery using MATLAB/SIMULINK, Prentice Hall, New Jersey, 1998.
[6] Safi, S. K., Acarnley, P. P., Jack, A. G., Analysis and simulation of the high-speed torque performance of brushless DC motor drives, IEE Proceedings – Electric Power Applications, 3 (142) (1995) 191–200.
[7] Sozer, Y., Torrey, D. A., Adaptive torque ripple control of permanent magnet brushless DC motors, Proceedings on the Applied Power Electronics Conference, Anaheim, USA, 1998, pp. 86–92.
[8] Yoon, Y., Kim, D., Lee, T., Choe, Y., Won, C., A low cost speed control system of PM brushless DC motor using 2 Hall-Ics, Proceedings of the International Conference on Mechatronics and Information Technology, Jecheon, Korea, 2003, pp. 150–155.

2

Finite element for non-stationary problems of viscoelastic orthotropic beams

M. Zajíček[a,*], V. Adámek[a], J. Dupal[a]

[a] *Faculty of Applied Sciences, University of West Bohemia, Univerzitní 22, 306 14 Plzeň, Czech Republic*

Abstract

The main aim of this work is to derive a finite beam element especially for solving of non-stationary problems of thin viscoelastic orthotropic beams. Presented approach combines the Timoshenko beam theory with the consideration of nonzero axial strain. Furthermore, the discrete Kelvin-Voight material model was employed for the description of beam viscoelastic material behaviour. The presented finite beam element was derived by means of the principle of virtual work. The beam deflection and the slope of the beam have been determined by the analytical and numerical (FEM) approach. These studies were made in detail on the simple supported beam subjected to the non-stationary transverse continuous loading described by the cosine function in space and by the Heaviside function in time domain. The study shows that beam deformations obtained by using derived finite element give a very good agreement with the analytical results.

Keywords: finite element, beam, Timoshenko theory, Kelvin-Voight material, non-stationary problems, analytical solution

1. Introduction

This work concerns the solution of the planar problem of a thin straight orthotropic viscoelastic beam subjected to a non-stationary loading. The main aim of this study is to derive the finite beam element based on the approximate Timoshenko beam theory and to compare analytical and numerical results for a particular beam problem. The purpose of this effort is to use such element for the effective numerical solution of beam-like structures inverse problems with minimal loss of solution accuracy. It is well known fact that inverse problems, e.g. material parameters identification etc., are usually very time-consuming and so the demand for effective computation is one of the most important (see e.g. [7]).

Many authors were concerned with analytical as well as finite element solutions of beam problems. The classical Euler-Bernoulli beam theory [2,4] is restricted for thin beams and does not include the effect of shear forces on the beam deformation. Due to this limitation, in 1921 Timoshenko developed a beam theory including the effect of the transverse shear deformation which is assumed constant across the thickness of the beam and depends on the shear correction factor (see e.g. [2,4]). Further, Chandrashekhara et al. [6] employed the analytical solution for the free vibration of laminated composite beams including the transverse (first-order) shear deformation effects and the rotary inertia. The solution procedure is applicable to arbitrary boundary conditions. Two higher order displacement based on shear deformation theories of free vibration analysis of laminated composite beams is carried out for example in [8].

---

*Corresponding author. e-mail: zajicek@kme.zcu.cz.

Finite elements have also been developed based on the Timoshenko theory as could be found in [3,10] or [12]. Most of these finite element models possess a two node-two degree of freedom structure because the requirements of the variational principle for the Timoshenko's displacement field are accepted. Davis et al. showed in [3] that a Timoshenko beam element converged to the exact solution of the elasticity equations for a simply supported beam provided that the correct value of the shear factor is used. Thomas and Abbas [10] presented for the first time a finite element model with nodal degrees of freedom which can satisfy all the forced and the natural boundary conditions of a Timoshenko beam. The mass and stiffness matrices of the element are derived from kinetic and strain energies by assigning polynomial expressions for total deflection and bending slope. Zienkiewicz and Taylor [12] give a summary of finite element models for the Euler-Bernoulli and the Timoshenko theory. A mathematical framework from which general problems may be formulated and solved using variational and Galerkin methods is presented. In addition, these authors consider problem of shape functions for situations in which the approximating functions (displacement and slope) are necessary C^0 and C^1 continuous. Zienkiewicz and Taylor also cover in some detail formulations for viscoelasticity, plasticity and viscoplasticity material models.

One can find several types of already derived finite beam elements which employed a higher order shear deformation theories. A second-order beam theory requiring two coefficients, one for cross-sectional warping and the second for transverse direct stress, was developed by Stephen and Levinson [9]. Heyliger and Reddy [5] used a higher order beam finite element for bending and vibration problems. In this formulation, the theory assumes a cubic variation of the in-plane displacement in thickness co-ordinate and a parabolic variation of the transverse shear stress across the thickness of the beam. Further this theory satisfies the zero shear strain conditions at the top and bottom surfaces of the beam and neglects the effect of the transverse normal strain. Subramanian [8] furthermore carried out two-node C^1 finite elements of eight degrees of freedom per node for the vibration problems of the laminated composite beams. Applied theories not only include the effect of transverse shear strain and normal strain but also satisfy free transverse shear strain/stress conditions on the top and bottom lateral surfaces of the beams.

This paper relates to the work [11] in which the analytical solution of static and free vibration problem of a uniform and linear elastic beam has been derived. In addition, results obtained in [1], where authors presented analytical and numerical solution of non-stationary vibrations of a thin viscoelastic beam, were utilized. In both mentioned works, beams were supposed as orthotropic and thin and their formulations were based on the Timoshenko beam theory.

2. Problem formulation

We generally consider a straight beam of length $2l$ consisting of n layers which are perfectly bonded. Let us number these layers from the lower to the upper face as shown in Fig. 1. The overall thickness of the laminated beam is h. Homogenous, orthotropic and linear viscoelastic (Kelvin-Voight model) material properties of layers are supposed. Each layer k is referred to by the x_3 coordinates of its lower face h_{k-1} and upper face h_k. The angle θ_k is the orientation of the kth layer (in directions L, T, T') with respect to the x_1-axis. The cross-section area of beams can have various geometries (with the width b) but must be uniform along the x_1-axis and symmetric to the x_3-axis. Furthermore, the general combination of lateral and axial loading may be applied but only bending and stretching in the $x_1 - x_3$ plane of symmetry can exist. This is the reason why only displacements $u_1(x_1, x_3, t)$ and $u_3(x_1, t)$ in the x_1 and x_3 directions,

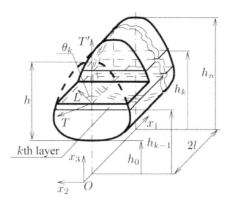

Fig. 1. A thin laminated beam with a symmetric cross-section

respectively, may be nonzero if the assumption is accepted that Poisson's effects may cause deformations only in the $x_1 - x_3$ plane.

The displacement fields for the first-order shear deformation theory are taken as

$$u_1(x_1, x_3, t) = u(x_1, t) + x_3\,\psi(x_1, t) \qquad \text{and} \qquad u_3(x_1, t) = w(x_1, t)\,, \tag{1}$$

where $u(x_1, t)$ is the displacement due to extension, $w(x_1, t)$ is the displacement due to bending and $\psi(x_1, t)$ represents rotation of the transverse normal referred to the plane $x_3 = 0$. Besides, $u(x_1, t)$ can be expressed in the form

$$u(x_1, t) = u_c(x_1, t) - z_c\psi(x_1, t) \tag{2}$$

with the help of displacement $u_c(x_1, t)$ in centroidal axis direction when the distance of this axis and x_1-axis is

$$z_c = \frac{B_{11}}{A_{11}}\,, \tag{3}$$

where B_{11} and A_{11} are stiffness parameters well-known in the laminate theory (see e.g. [11]). The nonzero strain-displacement relationships for presented theory are then given as

$$\varepsilon_{11}(x_1, x_3, t) = \frac{\partial u_1}{\partial x_1} = \frac{\partial u}{\partial x_1} + x_3\frac{\partial \psi}{\partial x_1}\,, \quad 2\,\varepsilon_{31}(x_1, t) = \frac{\partial u_3}{\partial x_1} + \frac{\partial u_1}{\partial x_3} = \frac{\partial w}{\partial x_1} + \psi = \gamma(x_1, t)\,. \tag{4}$$

To make following relations more transparent, the notation for strain and stress tensor components are reduced to $\varepsilon_{11} = \varepsilon_1, 2\,\varepsilon_{31} = \varepsilon_5, \sigma_{11} = \sigma_1$ and $\sigma_{31} = \sigma_5$. The stress-strain relationships for the kth layer is then, with respect of (4), taken as

$$\sigma_i^k = Q_{ii}^k\,\varepsilon_i + \widetilde{Q}_{ii}^k\,\dot{\varepsilon}_i \qquad \text{for} \qquad i = 1, 5\,, \tag{5}$$

where

$$Q_{11}^k = \frac{E_1}{1 - \mu_{12}\mu_{21}}\,, \qquad Q_{55}^k = G_{31}\,, \qquad \widetilde{Q}_{11}^k = \frac{\lambda_1}{1 - \nu_{12}\nu_{21}} \qquad \text{and} \qquad \widetilde{Q}_{55}^k = \eta_{31} \tag{6}$$

are the reduced material constants. It is obvious that the Young's modulus E_1 is equal to the longitudinal modulus E_L and the transverse modulus E_T for $\theta_k = 0$ and $\theta_k = 90$, respectively. It is similarly valid for the coefficient of normal viscosity λ_1. The other material constants (i.e. the Poisson's ratios μ_{12}, μ_{21} and ν_{12}, ν_{21} of elastic and viscous elements in material model, respectively, shear modulus G_{31} and the coefficient of shear viscosity η_{31}) in coordinate system x_1, x_2, x_3 are the same as in the coordinate system L, T, T' for both angles θ_k.

3. Finite element formulation

In this paper, the shape functions are formulated using weak form for linear elastic rods, see [12]. When the longitudinal and centroidal axes are identical, the displacement field can be restricted to bending and axial strains, eqs. (1) and (2). Then the solution of rod problem involves finding one-dimensional interpolations to approximate each of these functions appearing in the weak form.

Zienkiewicz and Taylor bring in [12] that the weight function for the axial deformation and the static problem must satisfy the exact solution of the adjoint homogeneous ordinary differential equation of the second order. Therefore the linear polynomial

$$u_c(x_1, t) = [1, x_1][a_0(t), a_1(t)]^T = \phi_1(x_1)c_1(t) \tag{7}$$

is selected to describe the centroidal deformation. If a static behaviour of the Euler-Bernoulli beam is considered, the exact interpolation for the transverse displacement can be also found. In order to obtain exact interelement nodal solution for ordinary differential equation, the interpolation function for the weight function must satisfy the exact solution of the adjoint homogeneous differential equation of the fourth order as mentioned in [12]. It is the reason why a polynomial of cubic order for Euler-Bernoulli beam shape function is to use, i.e.

$$w(x_1, t) = [1, x_1, x_1^2, x_1^3][a_2(t), a_3(t), a_4(t), a_5(t)]^T = \phi_2(x_1)c_2(t). \tag{8}$$

It could be also found in [12] that the proposed transverse and centroidal axis displacement approximations may be used in transient analysis as well, however then the solution is no longer be exact at the interelement nodes. Therefore vectors $c_1(t)$ and $c_2(t)$ in (7) and (8) are generally assumed time-dependent.

Implementation of the transverse shear strain contribution to the rotation angle about the coordinate axis x_2, i.e. the application of the Timoshenko theory, was made by the next way. We consider equations of motion

$$B_{11}u''(x_1) + D_{11}\psi''(x_1) - \alpha A_{55}\gamma(x_1) = 0, \tag{9}$$

see [11], where inertial forces were neglected and distributed forces per length on the beam faces are omitted. It means in consequence that the transverse force and transverse shear strain are constant along the element length. Inserting displacement approximations mentioned above into modified balance equations, we are able to determine the rotation angle including the transverse shear strain

$$\psi(x_1, t) = \gamma - \frac{\partial w}{\partial x_1} = [\phi_5 - \phi_2'(x_1)]\,c_2(t) = \phi_3(x_1)c_2(t) \tag{10}$$

with

$$\phi_5 = [0, 0, 0, -6\,D_T/(\alpha A_{11}A_{55})] \qquad \text{and} \qquad D_T = A_{11}D_{11} - B_{11}^2. \tag{11}$$

The parameter α is the shear correction factor and may be calculated as shown in [11]. The stiffness constants are defined as follows

$$(A_{11}, B_{11}, D_{11}) = \sum_{k=1}^{n} Q_{11}^k \int_{h_{k-1}}^{h_k} b(x_3)(1, x_3, x_3^2)\,dx_3, \qquad A_{55} = \sum_{k=1}^{n} Q_{55}^k \int_{h_{k-1}}^{h_k} b(x_3)\,dx_3. \tag{12}$$

Now we consider the division of the beam domain V into a set of disjoint subdomains V^e such that the sum over the element domains V^e is equal to V. Similarly the boundary is divided into subdomains. Because the finite beam element is developed by means of the principle of virtual work (see e.g. [12]), we obtain after some simplifications the relation

$$\sum_e \left\{ \sum_{j=1,5} \int_{V^e} \delta \varepsilon_j \, \sigma_j \, \mathrm{d}V + \sum_{i=1,3} \left[\int_{V^e} \delta u_i \left(\rho \, \ddot{u}_i - X_i \right) \mathrm{d}V - \int_{A^e_p} \delta u_i \, p_i \, \mathrm{d}A \right] \right\} = 0 \,, \qquad (13)$$

where A^e_p is a boundary segment on which tractions p_i are specified and where X_i are body force components. It follows from (13) that the definition of the local axis x_e within the domain V^e of each element is useful. Consequently, generalized deformations in element nodes may be given as

$$\boldsymbol{q}_1(t) = [u(0,t), u(l_e,t)]^T \qquad \text{and} \qquad \boldsymbol{q}_2(t) = [w(0,t), w(l_e,t), \psi(0,t), \psi(l_e,t)]^T, \qquad (14)$$

while the finite element is considered of length l_e. Using the approximate functions (7), (8) and (10) which are expressed in local coordinates leads to the system of equations

$$\begin{bmatrix} \boldsymbol{q}_1(t) \\ \boldsymbol{q}_2(t) \end{bmatrix} = \begin{bmatrix} \boldsymbol{S}_{11} & -z_c \boldsymbol{S}_{12} \\ \boldsymbol{0} & \boldsymbol{S}_{22} \end{bmatrix} \begin{bmatrix} \boldsymbol{c}_1(t) \\ \boldsymbol{c}_2(t) \end{bmatrix}, \quad \boldsymbol{S}_{11} = \begin{bmatrix} \phi_1(0) \\ \phi_1(l_e) \end{bmatrix}, \quad \boldsymbol{S}_{12} = \begin{bmatrix} \phi_3(0) \\ \phi_3(l_e) \end{bmatrix}, \quad \boldsymbol{S}_{22} = \begin{bmatrix} \phi_2(0) \\ \phi_2(l_e) \\ \phi_3(0) \\ \phi_3(l_e) \end{bmatrix}. \qquad (15)$$

Solving the equation (15), we get time-dependent functions

$$\boldsymbol{c}_1(t) = \boldsymbol{S}_{11}^{-1} \boldsymbol{q}_1(t) + z_c \boldsymbol{S}_{11}^{-1} \boldsymbol{S}_{12} \boldsymbol{S}_{22}^{-1} \boldsymbol{q}_2(t) \qquad \text{and} \qquad \boldsymbol{c}_2(t) = \boldsymbol{S}_{22}^{-1} \boldsymbol{q}_2(t) \,. \qquad (16)$$

Substituting this result into (7), (8) and (10), the finite element approximation for displacements could be rewritten as follows

$$\begin{aligned} u(x_e,t) &= \phi_1(x_e) \boldsymbol{S}_{11}^{-1} \boldsymbol{q}_1(t) + \phi_4(x_e) \boldsymbol{S}_{22}^{-1} \boldsymbol{q}_2(t) \,, \\ w(x_e,t) &= \phi_2(x_e) \boldsymbol{S}_{22}^{-1} \boldsymbol{q}_2(t) \,, \\ \psi(x_e,t) &= \phi_3(x_e) \boldsymbol{S}_{22}^{-1} \boldsymbol{q}_2(t) \end{aligned} \qquad (17)$$

with

$$\phi_4(x_e) = z_c \left[\phi_1(x_e) \boldsymbol{S}_{11}^{-1} \boldsymbol{S}_{12} - \phi_3(x_e) \right] \,. \qquad (18)$$

To derive mass, stiffness and damping matrices of the finite element using the principle of virtual work, eqs. (4), (5) and (17) are substituted into (13). If the matrix $\boldsymbol{I}_{ij}^{kl}$ is defined as

$$\boldsymbol{I}_{ij}^{kl} = \int_0^{l_e} \frac{\partial^i \phi_k^T}{\partial x_e^i} \frac{\partial^j \phi_l}{\partial x_e^j} \, \mathrm{d}x_e \,, \qquad (19)$$

we can write

$$\sum_{i=1,3} \int_{V^e} \rho \, \delta u_i \, \ddot{u}_i \, \mathrm{d}V = [\delta \boldsymbol{q}_1^T(t), \delta \boldsymbol{q}_2^T(t)] \begin{bmatrix} \boldsymbol{M}_{e11} & \boldsymbol{M}_{e12} \\ \boldsymbol{M}_{e12}^T & \boldsymbol{M}_{e22} \end{bmatrix} \begin{bmatrix} \ddot{\boldsymbol{q}}_1(t) \\ \ddot{\boldsymbol{q}}_2(t) \end{bmatrix} = \delta \boldsymbol{q}_e^T(t) \, \boldsymbol{M}_e \, \ddot{\boldsymbol{q}}_e(t) \,, \qquad (20)$$

$$\sum_{j=1,5} \int_{V^e} \delta \varepsilon_j \, \sigma_j \, \mathrm{d}V = [\delta \boldsymbol{q}_1^T(t), \delta \boldsymbol{q}_2^T(t)] \left\{ \begin{bmatrix} \boldsymbol{K}_{e11} & \boldsymbol{K}_{e12} \\ \boldsymbol{K}_{e12}^T & \boldsymbol{K}_{e22} \end{bmatrix} \begin{bmatrix} \boldsymbol{q}_1(t) \\ \boldsymbol{q}_2(t) \end{bmatrix} + \begin{bmatrix} \boldsymbol{B}_{e11} & \boldsymbol{B}_{e12} \\ \boldsymbol{B}_{e12}^T & \boldsymbol{B}_{e22} \end{bmatrix} \begin{bmatrix} \dot{\boldsymbol{q}}_1(t) \\ \dot{\boldsymbol{q}}_2(t) \end{bmatrix} \right\} =$$
$$\delta \boldsymbol{q}_e^T(t) \left\{ \boldsymbol{K}_e \, \boldsymbol{q}_e(t) + \boldsymbol{B}_e \, \dot{\boldsymbol{q}}_e(t) \right\} \,, \qquad (21)$$

where

$$M_{e11} = S_{11}^{-T}\rho_{11}I_{00}^{11}S_{11}^{-1},$$
$$M_{e12} = S_{11}^{-T}\left[\rho_{11}I_{00}^{14} + R_{11}I_{00}^{13}\right]S_{22}^{-1},$$
$$M_{e22} = S_{22}^{-T}\left[\rho_{11}\left(I_{00}^{22} + I_{00}^{44}\right) + R_{11}\left(I_{00}^{34} + I_{00}^{43}\right) + I_{11}I_{00}^{33}\right]S_{22}^{-1}, \tag{22}$$
$$K_{e11} = S_{11}^{-T}A_{11}I_{11}^{11}S_{11}^{-1},$$
$$K_{e12} = S_{11}^{-T}\left[A_{11}I_{11}^{14} + B_{11}I_{11}^{13}\right]S_{22}^{-1},$$
$$K_{e22} = S_{22}^{-T}\left[A_{11}I_{11}^{44} + \alpha A_{55}I_{00}^{55} + B_{11}\left(I_{11}^{34} + I_{11}^{43}\right) + D_{11}I_{11}^{33}\right]S_{22}^{-1}, \tag{23}$$
$$B_{e11} = S_{11}^{-T}\widetilde{A}_{11}I_{11}^{11}S_{11}^{-1},$$
$$B_{e12} = S_{11}^{-T}\left[\widetilde{A}_{11}I_{11}^{14} + \widetilde{B}_{11}I_{11}^{13}\right]S_{22}^{-1},$$
$$B_{e22} = S_{22}^{-T}\left[\widetilde{A}_{11}I_{11}^{44} + \alpha\widetilde{A}_{55}I_{00}^{55} + \widetilde{B}_{11}\left(I_{11}^{34} + I_{11}^{43}\right) + \widetilde{D}_{11}I_{11}^{33}\right]S_{22}^{-1}. \tag{24}$$

In these relations, we can find other parameters except A_{11}, B_{11}, D_{11} and A_{55} defined by (12). Then, the mass moment of inertia terms are given as

$$(\rho_{11}, R_{11}, I_{11}) = \sum_{k=1}^{n}\rho^k\int_{h_{k-1}}^{h_k}b(x_3)(1, x_3, x_3^2)\,\mathrm{d}x_3, \tag{25}$$

where ρ^k is the mass density of the kth material layer, and parameters of damping are taken as

$$(\widetilde{A}_{11}, \widetilde{B}_{11}, \widetilde{D}_{11}) = \sum_{k=1}^{n}\widetilde{Q}_{11}^k\int_{h_{k-1}}^{h_k}b(x_3)(1, x_3, x_3^2)\,\mathrm{d}x_3, \qquad \widetilde{A}_{55} = \sum_{k=1}^{n}\widetilde{Q}_{55}^k\int_{h_{k-1}}^{h_k}b(x_3)\,\mathrm{d}x_3. \tag{26}$$

As obvious from (13), each finite element is generally subjected to body forces and tractions. These loadings may be time dependent and therefore can be rewritten in the form

$$\sum_{i=1,3}\left(\int_{V^e}\delta u_i X_i\,\mathrm{d}V + \int_{A_p^e}\delta u_i p_i\,\mathrm{d}A\right) = \delta q_e^T(t)f_e(t), \tag{27}$$

where $f_e(t)$ is the vector of element external loading.

Expressing the vector q_e using the corresponding localization matrix J_e for each element by the relation

$$q_e(t) = J_e q(t) \tag{28}$$

and substituting (20), (21) and (27) into (13), the solution of a problem by a finite element method leads to differential equations of the form

$$M\ddot{q}(t) + B\dot{q}(t) + Kq(t) = f(t) \qquad \text{for} \qquad \delta q^T(t) \neq 0, \tag{29}$$

where

$$M = \sum_e J_e^T M_e J_e, \qquad B = \sum_e J_e^T B_e J_e, \qquad K = \sum_e J_e^T K_e J_e \tag{30}$$

are mass, damping and stiffness matrices, respectively, and

$$f(t) = \sum_e J_e^T f_e(t) \tag{31}$$

is time dependent vector of external loading. The vector $q(t)$ represents generalized displacements in nodes that can be obtained by solving the system (29) with the help of some numerical integration method.

4. Analytical solution used for finite element verification

As mentioned above, the verification of derived finite element was performed using the analytical solution of a particular beam problem. The analytical solution of non-stationary vibration of a simply supported thin viscoelastic beam presented in [1] was used for this purpose. The geometry of this problem is depicted in Fig. 2. A thin beam of length $2l$ and with rectangular cross-section $b \times h$ is on its upper surface excited by a non-uniform tranverse external loading that is nonzero only on the region of length $2d$. The spatio-temporal function describing this loading was assumed in the form

$$q(x_1, t) = \sigma_a \, b \, \cos \frac{\pi (x_1 - l)}{2d} H(t) \qquad \text{for} \qquad x_1 \in \langle l - d, l + d \rangle, \tag{32}$$

where σ_a represents the excitation amplitude and $H(t)$ denotes the Heaviside function in time.

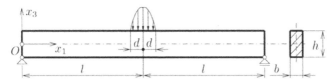

Fig. 2. Geometry of a simple supported beam with applied loading

Using the Timoshenko beam theory and solving the resulting system of two partial integro-differential motion equations by the method of integral transforms, the beam deflection function $w(x_1, t)$ and the beam slope function $\psi(x_1, t)$ can be expressed as infinite sums in the form [1]

$$w(x_1, t) = \sum_{n=1}^{\infty} C(n) \sin(\omega_n x_1) \, \mathcal{L}^{-1} \left\{ \frac{H_4(n, p)}{p H_6(n, p)} \right\},$$

$$\psi(x_1, t) = \sum_{n=1}^{\infty} C(n) \cos(\omega_n x_1) \, \mathcal{L}^{-1} \left\{ \frac{H_2(n, p)}{p H_6(n, p)} \right\}, \tag{33}$$

where the operator $\mathcal{L}^{-1}$ denotes the inverse Laplace transform and the real function $C(n)$ related to the applied loading is defined as

$$C(n) = 4 \, \sigma_a \, d \, b \, \frac{\cos\left(\frac{\pi n d}{2l}\right) \sin\left(\frac{\pi n}{2}\right)}{\pi \, l \left[1 - \left(\frac{n d}{l}\right)^2\right]}. \tag{34}$$

The complex functions $H_i(n, p)$ $(i = 2, 4, 6)$ in (33) are introduced by relations

$$H_2(n, p) = 12 \, \alpha \, \omega_n \, G^*(p), \qquad H_4(n, p) = h^2 \left(\rho \, p^2 + \omega_n^2 \, E^*(p)\right) + 12 \, \alpha \, G^*(p),$$

$$H_6(n, p) = b \, h \left[\left(\rho \, p^2 + \alpha \, \omega_n^2 \, G^*(p)\right) H_4(n, p) - \frac{1}{12} H_2(n, p)^2\right], \tag{35}$$

where $\omega_n = \frac{n\pi}{2l}$. Viscoelastic material properties of the beam are described by the complex moduli $E^*(p)$ a $G^*(p)$ which, in the case of generalized standard viscoelastic solid model with N parallel Maxwell elements, can be expressed as

$$E^*(p) = \sum_{k=0}^{N} E_{1_k} - \sum_{k=1}^{N} \frac{E_{1_k}^2}{\lambda_{1_k} p + E_{1_k}}, \qquad G^*(p) = \sum_{k=0}^{N} G_{31_k} - \sum_{k=1}^{N} \frac{G_{31_k}^2}{\eta_{31_k} p + G_{31_k}}. \tag{36}$$

The analytical solution for the Kelvin-Voight material model can be then simply obtained by the assumption $N = 1$ and $E_{1_1} \to \infty$.

5. Numerical results and discussion

In this section, the comparison of analytical and numerical results obtained by different integration methods and using two beam theories is given. Moreover, the verification of the analytical approach is performed with the help of FE solution using the software MSC.Marc. All computations were made for the problem depicted in Fig. 2 for the time interval $t \in \langle 0, 200 \rangle \, \mu$s under the assumption $\sigma_a = 1$ MPa and $2 \, d = 4$ mm. Furthermore, the geometric dimensions were taken as $l = 50$ mm, $h = 10$ mm and $b = 5$ mm. The value of the shear correction factor corresponds to arbitrary rectangular cross-section, i.e. $\alpha = 0.833$.

The parameters describing material behaviour of the beam studied were assumed as follows: $\rho = 2.1 \cdot 10^3 \, \text{kg m}^{-3}$, $E_1 = 39$ GPa, $G_{31} = 3.8$ GPa, $\lambda_1 = 35$ kPa s. Elastic and viscous Poisson's ratios were the same and equal to 0.28. It should be emphasized that the used material parameters do not correspond to a real material because it is nearly impossible to find values of all needed parameters for orthotropic material in literature. That is why some properties of unidirectional composite material E-glass/epoxy were used for the estimation of required parameters.

Points with coordinate $x_1 \in \{52, 60, 70\}$ mm, where the deflection and the slope of the beam were calculated, have been selected as points of interest. This choice allowed us to compare analytical and numerical solutions both in the vicinity of applied loading (the first point was identical to the boundary of loaded domain) and far from it. The computation of all presented problems have been done on a PC Pentium 4 with CPU 2.99 GHz and with RAM 3 GB.

5.1. Results given by MSC.Marc software

The numerical solution obtained using FE software MSC.Marc served for the verification of the analytical solution quality whereas the problem was solved as the problem of plane stress. Because the condition of symmetry were accepted, only the half of the beam geometry was modeled. The beam mesh consisted of 12500 regular four-node isoparametric elements with linear approximations. The basic element size 0.2×0.2 mm was chosen according to the work [1] and only the elements near the area of excitation were once refined to the half of their original size to reach better representation of applied loading. In addition to the zero axial displacement defined on the axis of symmetry, the boundary condition representing the beam support was prescribed in the first node at the bottom beam edge. Material properties of the beam were represented by the discrete viscoelastic model which is implemented in the used software.

The integration in time domain was performed by the Newmark method with a constant acceleration and with integration step $4 \cdot 10^{-8}$ s. This value was determined with respect to the maximum stable integration step of the explicit scheme of central differences [1]. This fact together with maximum time of interest led to very time-consuming computation which took about 4.7 hours.

As it is shown in Fig. 3, the comparison of the history plots of deflection w in points of interest between analytical (A.S. lines) and numerical (FEM lines) approach has been done. FEM results were investigated on the upper beam edge. It is obvious from Fig. 3(a) that both solution correspond each other quite well in all points studied. Moreover, the detailed view in Fig. 3(b) shows that some oscillations of the numerical solution occur. These phenomena are particularly clear for $x_1 = 52$ mm. The reason of these oscillations lies in the two-dimensionality of the FE numerical model. The obtained results show that the presented analytical solution may be employed for studying quality of derived finite element.

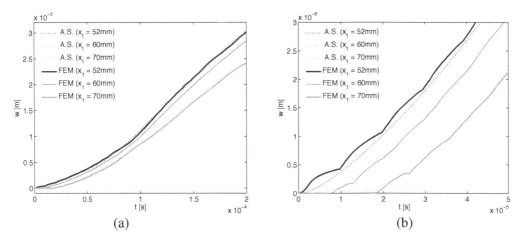

Fig. 3. History plot of beam deflection in points of interest (a) and detailed view (b). Analytical approach versus results given using software MSC.Marc

5.2. Results given by FE solution based on Euler-Bernoulli theory

In this part, numerical results computed using the finite elements developed based on the Euler-Bernoulli theory were compared with the analytical solution. For this purpose the FE model A built with 50 elements and FE model B built with 500 elements were presented. In both cases, finite elements were taken with the constant length along longitudinal beam axis. Euler-Bernoulli beam elements needed in these analyses can be simply obtained by omitting the vector ϕ_5 in (11). Therefore, the derived element in this part was utilized with some simplifications. Only essential boundary conditions, i.e. zero displacements at the end nodes of mesh according to Fig. 2, were applied in cases A and B.

Central difference method with integration step $1.25 \cdot 10^{-7}$ s (problem A) and Newmark method with a constant acceleration and with integration step $1 \cdot 10^{-7}$ s (problem B) were used for numerical integration of the system (29). Time steps were chosen with respect to the elements length and to the phase velocity of longitudinal wave in a thin rod. That is why the computation times of A and B problems were 0.8 s and 56 s, respectively, when the whole time interval $t \in \langle 0, 200 \rangle$ μs were taken into account.

It is observed from Fig. 4 given for the beam deflection (a) and the slope of beam (b) that correspondence between analytical and numerical results is rather bad. We can find cumulative difference in the beam deflection at all points of interest with increasing time. The curves describing the beam slope have even different form. Consequently, FE model contained 500 elements and solved with the help of central difference method was made but results were similar to them in the case B.

5.3. Results given by FE solution based on Timoshenko theory

In what follows, numerical tests using the Timoshenko beam theory are performed to confirm the quality of the finite element derived. Therefore, deformations of simply supported beam for variants 1 to 5 shown in Table 1 were studied. As can be seen for individual models in this table, meshes of solved problems consisted of 50 or 500 elements of equal lengths. Further, boundary conditions and numerical integration methods were used, as well as in cases of finite element models based on the Euler-Bernoulli theory.

It is clear from Table 1 that beam deformations obtained by FEM (in time $t = 200\,\mu$s) are slightly different from their exact values (errors are less than $0.2\,\%$). Similarly, good agreements

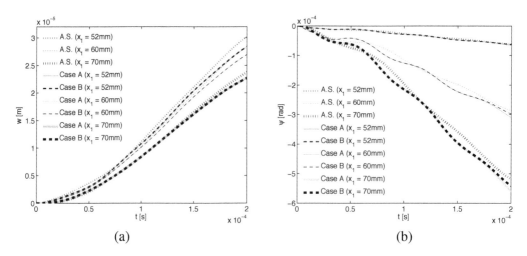

Fig. 4. History plot of beam deflection (a) and slope of beam (b) in points of interest. Analytical approach versus FE solution based on Euler-Bernoulli theory

Table 1. Comparison of numerical and analytical results in time $200 \, \mu s$

Model	Method	Elem.	Integration	CPU time	Error [%] $(T = 200 \, \mu s)$					
					w			ψ		
m			step [s]	[s]	x_1 [mm]					
					52	60	70	52	60	70
1	Cent. diff.	500	2.50e−8	216	0.013	0.013	0.005	0.050	0.019	0.011
2	Newmark	500	1.00e−7	54	0.001	0.001	−0.008	0.040	0.008	−0.001
3	Cent. diff.	50	2.50e−7	0.58	0.194	0.185	0.167	0.110	0.079	0.069
4	Newmark	50	1.00e−6	0.23	0.074	0.063	0.043	0.005	−0.031	−0.047
5	Newmark	50	2.50e−7	0.61	0.074	0.063	0.042	0.010	−0.026	−0.045

in absolute values of deformations are found in all points of investigated time interval. In order to make the mutual comparison of results accuracy over all cases $(m = 1, \ldots, 5)$, following parameters have been defined:

$$w^*(x_1, t, T)\big|_r^m = \frac{w_m(x_1, t) - w_{\text{A.S.}}(x_1, t)}{w_r(x_1, T) - w_{\text{A.S.}}(x_1, T)},$$

$$\psi^*(x_1, t, T)\big|_r^m = \frac{\psi_m(x_1, t) - \psi_{\text{A.S.}}(x_1, t)}{\psi_r(x_1, T) - \psi_{\text{A.S.}}(x_1, T)}, \tag{37}$$

where $w_m(x_1, t)$ and $w_{\text{A.S.}}(x_1, t)$ is beam deflection calculated from mth problem and analytical solution, respectively. In the analogous way, the notation of beam slope $\psi_m(x_1, t)$ and $\psi_{\text{A.S.}}(x_1, t)$ is employed. It is shown in Table 1 that the minimum difference were found for model $m = 2$. Consequently, the parameter $r = 2$ were taken for the calculation of $w^*\big|_r^m$ and $\psi^*\big|_r^m$ in (37). Fig. 5 shows history plots of these parameters in selected points of interest within the time interval $\langle 0, T \rangle$ where $T = 200 \, \mu s$. It is obvious from this figure that both solutions with a fine-mesh discretization $(m = 1, 2)$ give very good results over the whole time interval but the best results are explicitly found for $m = 2$. Furthermore, comparing results of all studied variants (mainly the deflections in Fig. 5(a) and Fig. 5(b)) it can be said that the better accuracy with

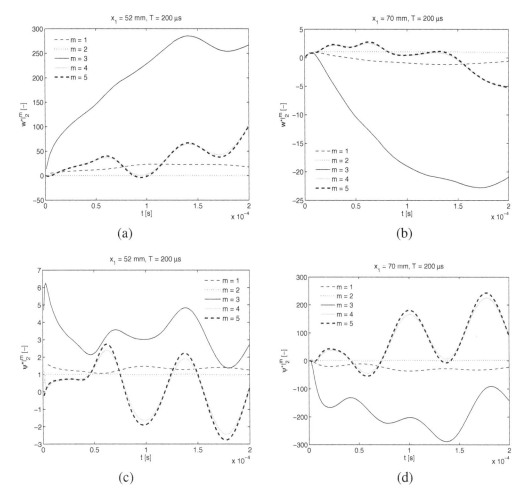

Fig. 5. History plot of parameters $w^*|_r^m$ and $\psi|_r^m$ for $r = 2$ and $T = 200\,\mu s$ in some points of interest. FE solution based on Timoshenko theory

respect to results of the model 2 can be achieved by the use of the Newmark integration method. The history plots for point of interest $x_1 = 60\,\mathrm{mm}$ were analysed as well but changing curves characters were similar to the case $x_1 = 70\,\mathrm{mm}$. Therefore these curves are not presented in this work.

6. Conclusion

The finite beam element for solving orthotropic viscoelastic problems was derived based on the Timoshenko theory using the principle of virtual work. The discrete Kelvin-Voight model was used for the description of its material behaviour. The validation of this element was made with the help of analytical solution and of the FE system MSC.Marc on the problem of simply supported beam subjected to a transverse non-stationary loading. Concretely, the time distribution of the beam deflection and the slope of the beam were compared in time interval 0 to $200\,\mu s$ and in three specific points ($x_1 \in \{52, 60, 70\}$ mm), one of which was situated in the vicinity of the loading applied. The beam deformations were calculated using the derived element, whereas the Euler-Bernoulli theory and the Timoshenko theory were taken into account. The meshes of numerical models consisted of 50 or 500 elements and the Newmark method or the method of central differences were used for the integration in time domain. Based on the computations performed, one can say that the developed finite element gives very accurate results and its usage is connected with significantly lower CPU-time than in the case of analytical approach. With respect to this fact, this element seems to be suitable for the effective solution of inverse problems.

Acknowledgements

This work has been supported by the research project MSM 4977751303 of the Ministry of Education of the Czech Republic.

References

[1] Adámek, V., Valeš, F., A viscoelastic orthotropic beam subjected to general transverse loading, Applied and Computational Mechanics 2 (2) (2008) 215–226.

[2] Altenbach, H., Altenbach, J., Kissing, W., Mechanics of composite structural elements, Springer, Berlin, 2004.

[3] Davis, R., Henshell, R. D., Wanburton, G. B., A Timoshenko beam element. Journal of Sound and Vibration (22) (1972) 475–487.

[4] Graff, K. F., Wave motion in elastic solids, Clarendon Press, Oxford, 1975.

[5] Heyliger, P. R., Reddy, J. N., A high order beam finite element for bending and vibration problems. Journal of Sound and Vibration (126) (1988) 309–326.

[6] Chandrashekhara, K., Krishnamurthy, K., Roy, S., Free vibration of composite beams including rotary inertia and shear deformation. Composite Structures (14) (1990) 269–279.

[7] Liu, G. R., Xi, Z. C., Elastic waves in anisotropic laminates, CRC Press, Boca Raton, 2002.

[8] Subramanian, P., Dynamic analysis of laminated composite beams using higher order theories and finite elements. Composite Structures (73) (2006) 342–353.

[9] Stephan, N. G., Levinson, M. A., A second-order beam theory. Journal of Sound and Vibration (67) (1979) 293–305.

[10] Thomas, J., Abbas, B. A. H., Finite element model for dynamic analysis of Timoshenko beams. Journal of Sound and Vibration (41) (1975) 291–299.

[11] Zajíček, M., Linear elastic analysis of thin laminated beams with uniform and symmetric cross-section, Applied and Computational Mechanics 2 (2) (2008) 397–408.

[12] Zienkiewicz, O. C., Taylor, R. L., Finite element method for solid and structural mechanics, Elsevier, Oxford, 6th edition, 2005.

3

An evaluation of the stress intensity factor in functionally graded materials

M. Ševčík[a,b,*], P. Hutař[a], L. Náhlík[a,b], Z. Knésl[a]

[a]Institute of Physics of Materials, Czech Academy of Sciences, Žižkova 22, 616 62 Brno, Czech Republic
[b]Institute of Solid Mechanics, Mechatronics and Biomechanics, Brno University of Technology, Technická 2, 616 69 Brno, Czech Republic

Abstract

Functionally graded materials (FGM) are characterised by variations in their material properties in terms of their geometry. They are often used as a coating for interfacial zones to protect the basic material against thermally or mechanically induced stresses. FGM can be also produced by technological process for example butt-welding of polymer pipes. This work is focused on a numerical estimation of the stress intensity factor for cracks propagating through FGM structure. The main difficulty of the FE model creation is the accurate description of continual changes in mechanical properties. An analysis of the FGM layer bonded from both sides with different homogenous materials was performed to study the influence of material property distribution. The thickness effect of the FGM layer is also discussed. All analyses are simulated as a 2D problem of an edge cracked specimen. In this paper, the above effects are quantified and conclusions concerning the applicability of the proposed model are discussed.

Keywords: functionally graded material, linear elastic fracture mechanics, discretization methodology, power-law material change

1. Introduction

Functionally graded materials (FGM) are composites where the composition varies from place to place in order to effect the best performance of the structure. The development of FGM has demonstrated its possible uses in a wide range of thermal and structural applications such as thermal barrier coatings, corrosion and wear resistant coatings and metal/ceramic joining. Mechanical properties gradation offers ways of optimizing structure and achieving high performance and material efficiency. At the same time, this optimization can result in numerous mechanical problems including estimation of effective properties and crack propagation behaviors in the final structure [3].

Functionally graded materials commonly occur in nature. The human body contains many examples of complex FGM parts, such as bones or teeth. Another example of naturally occurring FGM is in bamboo [16], see Fig. 1. The cross section of bamboo resembles a fibre-reinforced composite material with continuous change of fibre density. This configuration leads to a continuous change in material properties which is, in fact, also the philosophy of modern FGM materials.

Approaches towards how to study fracture behaviors of FGM structures are available in the literature. Many of them focus on analytical solutions, see [4, 5, 9]. The comprehensive

---

*Corresponding author. e-mail: sevcik@ipm.cz.

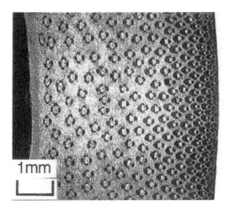

Fig. 1. Cross section of bamboo culm [16]

review concerning with the use of weight method for studying the crack propagation in FGM was published in [2]. The FGM structures have also been studied by numerical approaches such as the extended finite element method (X-FEM) [6, 7] or the finite element method (FEM) [3, 13, 14, 15].

Recent numerical simulations focus mainly on two dimensional (2D) analyses of crack propagation in FGM structures. This is due to difficulties occurring during FE model creation. The aim of this work is to develop a suitable discretization method for definition of the mechanical properties of the FGM applicable in common FE codes. This method should have greater accuracy and a lower computational time requirement which will prove beneficial during three dimensional analyses (3D) of much more complicated problems.

As the simulations performed in this work take into account the continual change of Young's modulus the linear elastic fracture mechanics approach is used to study crack behavior.

2. Estimation of stress intensity factor

In the Williams expansion (see e.g. [1]) for a linear elastic crack-tip stress field, the stress intensity factor corresponds to the first singular term:

$$\sigma_{ij} = \frac{K_I}{\sqrt{2\pi r}} f_{ij}(\theta) + T\delta_{1i}\delta_{1j} \dots \tag{1}$$

Here K_I is a stress intensity factor (only the normal mode of loading is considered), T is a T-stress, and $f_{ij}(\theta)$ is a known function of the polar angle θ. The stress intensity factor is widely used in many applications and the quality of this solution has been confirmed by numerous experimental results.

The validation of the use of the stress intensity factor in FGM layers has been achieved by many authors, e.g. [8] or [12]. Generally speaking, there are numerous ways how to estimate the stress intensity factor. One of the most common is by using quarter-point singular elements around the crack tip. This distortion of the FE mesh causes singularity $r^{-1/2}$. Due to the distortion, the stress field at the vicinity of the crack tip is better described and a very fine mesh is not necessary.

3. Coupled model of FGM

3.1. Material properties

In reality, the FGM structure is often connected with other materials. FGM layers often serve as a connection between two homogenous materials. In such cases, we need to simulate a complete configuration consistent with homogeneous parts connected to FGM, see [11]. The corresponding model of this set is shown in Fig. 2, where M1 indicates material number 1 (e.g. Al_2O_3) and M2 indicates material number 2 (e.g. Ni).

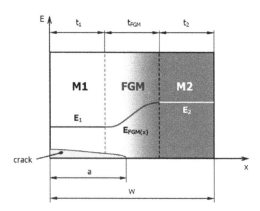

Fig. 2. Crack in the FGM layer

The Young's modulus E is constant in both materials M1 and M2. Due to the fact that only elastic material properties are used for presented simulations the linear elastic fracture mechanics approach is used for description of the crack behavior. In the FGM layer, the continuous change of E_{FGM} from E_1 (corresponding to material M1) to E_2 (material M2) is described as a function of the coordinates. In this general numerical study, the exact values of material properties are not necessarily known. The variable material property studied here is a ratio of Young's modulus E_2/E_1. In the following it is assumed that $E_1 = 1\,000$ MPa. The value of Young's modulus of material M2 is given by the ratio E_2/E_1. In this work, two values of the ratio E_2/E_1 are studied, namely $E_2/E_1 = 0.1$ – the crack spreads to the softer material and $E_2/E_1 = 10$ – the crack spreads to the tougher material. To describe the material properties distribution the following functions are commonly used in the literature [3, 10, 17]:

a) exponential function

$$f_{FGM}(x) = f_1 e^{[\beta(x/w)]},$$
$$\beta = \ln(f_2/f_1),$$

(2)

where f_i, $i = 1, 2$, is an arbitrary material property specified for material 1 or 2, β is constant of non-homogeneity, x is Cartesian coordinate, w is the width of the specimen, see Fig. 2

b) power-law function

$$f_{FGM}(x) = f_1 + (f_2 - f_1)(x/w)^g,$$

(3)

where g is a constant describing the gradient of material changes

c) double power-law function

$$h_1(x) = \tfrac{1}{2}\left(\tfrac{x}{w/2}\right)^p \qquad 0 \leq x \leq w/2$$

$$h_2(x) = 1 - \tfrac{1}{2}\left(\tfrac{w-x}{w/2}\right)^p \qquad w/2 \leq x \leq w, \tag{4}$$

$$f_{FGM}(x) = h_1(x)f_2 + [1 - h_1(x)]f_1 \qquad 0 \leq x \leq w/2$$

$$f_{FGM}(x) = h_2(x)f_2 + [1 - h_2(x)]f_1 \qquad w/2 \leq x \leq w, \tag{5}$$

where p is a degree of the polynomial, $h(x)$ determines volume fraction of components

d) rule of mixture

$$f_{FGM}(x) = v_f(x)E_2 + [1 - v_f(x)]E_1, \tag{6}$$

where $v_f(x)$ is local volume fraction of material.

The exponential change is often used for analytical approaches because of its easy numerical manipulation. The linear shape can be determined if $g = 1$ in case of the power-law function or for $p = 1$ in the case of a double-power law function. The rule of mixture is useful for those cases where the volume fraction of the material is known. In this study material properties distribution is determined by the power-law function (b) for all analyses.

3.2. Discretization of the material properties

In this paper, the new type of discretization has been developed. The structure is divided into a certain number of strips of different thickness t^i depending on the material properties distribution. The idea of this approach is shown in Fig. 3. The parameter controlling the number of the strips is a step factor s. The choice of its optimal values is studied in this paper.

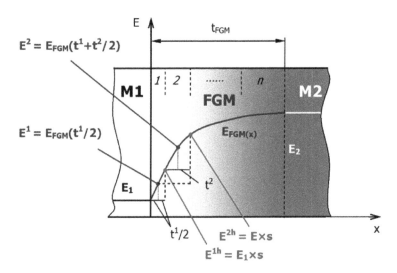

Fig. 3. The proposed discretization method

The nonhomogenous discretization of the continuous change of the material properties leads to n strips, which are perpendicular to the crack propagation direction. The advantage of this approach is a better description of the material property changes in positions where a strong gradient is presented. In these positions, more strips with smaller thickness are created and thus better accuracy is achieved. The dependence between a step factor s and the number of strips n corresponds to the geometry studied and the material changes used in the FGM layer is given in Table 1.

Table 1. Number of the strips n as a function of the step factor s

s	n	
	$E_2/E_1 = 10$	$E_2/E_1 = 0.1$
1.05	46	44
1.1	24	22
1.2	13	11
1.5	6	6
2	4	4
5	2	2

3.3. Boundary conditions

The boundary conditions (see Fig. 4) correspond to displacement loading. The top nodes are loaded by tensile stress $\sigma = 1$ MPa and the nodes are coupled so that the vertical displacements are identical. Because of the model symmetry, only one half is modelled. The problem is considered as 2D under plane strain conditions.

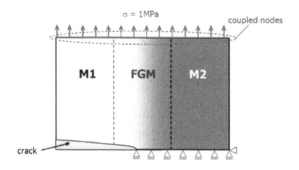

Fig. 4. Boundary conditions of coupled model of FGM – displacement loading

4. Results and discussion

4.1. Estimation of the step factor value

The aim of the first analysis was to find a suitable value for the step factor s. This parameter controls the number of the strips and the accuracy of the results. For that reason, an analysis of the influence of s was performed. The values of $s = 1.05$–5 and the ratios $E_2/E_1 = 0.1$ and $E_2/E_1 = 10$ were considered and analysed. The change in Young's modulus across the FGM layer was assumed to be linear. The results are presented in Fig. 5 and 6.

This analysis gives the estimation of the strip numbers for describing the FGM layer. For $n = 44$ and $n = 22$ the values of stress intensity factor are almost identical, the difference is less than 0.5 %. In case of the number $n = 11$ the difference is less then 1.5 %, which is a

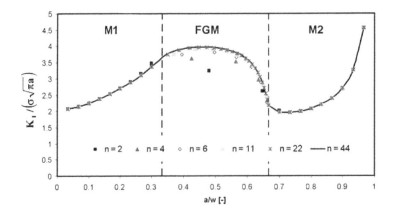

Fig. 5. Sensitivity of the stress intensity factor to a number of the areas – $E_2/E_1 = 0.1$

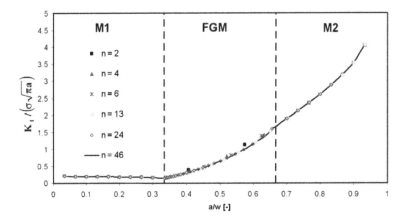

Fig. 6. Sensitivity of the stress intensity factor to a number of areas – $E_2/E_1 = 10$

good approximation. However, for $n = 6$, $n = 4$ and $n = 2$ the results are incorrect because the values of the stress intensity factor are rapidly underestimated. A similar analysis to the previous was performed but the ratio $E_2/E_1 = 10$ was assumed, see Fig. 6.

In comparison with the previous analysis ($E_2/E_1 = 0.1$), the stress intensity factor reaches significantly different values for $E_2/E_1 = 10$. The stress intensity factor is considerably lower in case of $E_2/E_1 = 10$. Another event is that the stress intensity factor is for ratio $E_2/E_1 = 10$ almost constant in material M1. This is caused by stiffer material in front of the crack. However, for the ratios $E_2/E_1 > 1$ the number of the strips in FGM layer can be very low. For example, the maximal difference between results for $n = 46$ and $n = 2$ was approximately 15 %. To conclude usage of 13–24 strips (corresponding to value of step factor $s = 1.2$–1.1) is in this case sufficient for correct simulation of the stress intensity factor in the FGM layers.

4.2. Influence of graded material properties on fracture parameters

Knowledge of the material properties distribution is necessary for simulating FGM of real structures. In general, the material here is described by a power function (see eq. 3). The distributions are described by the parameter g – gradient index, see Figs. 7a) and 7b). The initial size of the defect corresponded to the ratio $a/w = 0.03$. The crack propagates in material M1 in 9 steps until the crack tip reaches the FGM layer.

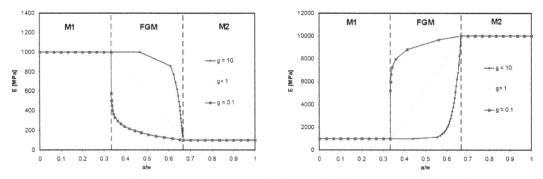

Fig. 7. Studied Young's modulus distributions a) $E_2/E_1 = 0.1$ b) $E_2/E_1 = 10$

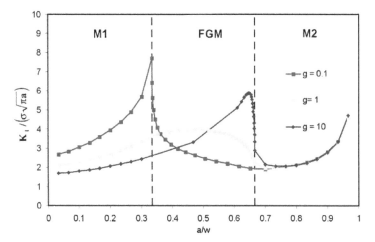

Fig. 8. Stress intensity factors $E_2/E_1 = 0.1$, displacement loading

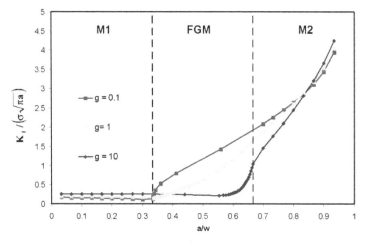

Fig. 9. Stress intensity factors $E_2/E_1 = 10$, displacement loading

Then the material properties of the FGM layer are divided by the discretization described into $n = 20$ strips in the case of $E_2/E_1 = 0.1$ and into $n = 24$ strips in the case of $E_2/E_1 = 10$. The results of the analyses are shown in Figs. 8 and 9. Significant differences in stress intensity factor values are evident. For the FGM layer with a prompt decrease in Young's modulus ($g = 0.1$) the dependence of the stress intensity factor looks similar to the crack penetrating

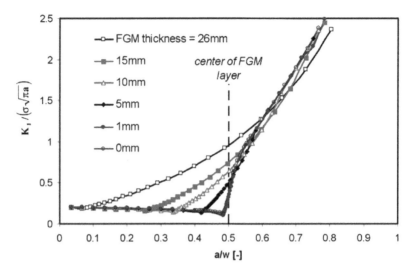

Fig. 10. Influence of the thickness of the FGM layer, $E_2/E_1 = 10$, displacement loading

the sharp interface. A very rapid decrease in Young's modulus appears immediately after the interface which caused the increase of the stress intensity factor. However, in the FGM layer the stress intensity factor rapidly decreases and starts to grow as far as material M2. A similar situation occurs for $g = 10$ where a strong change in the stress intensity factor value occurs near the interface of the FGM and material M2. The peak is nearly at the interface but the analysis showed a decrease of the stress intensity factor in the region in front of the FGM/M2 interface, see Fig. 8. In the case of linear change ($g = 1$) there is continuous smooth dependence. This is due to a gradual change in the Young's modulus.

As mentioned earlier, the stress intensity factor is almost constant for ratio $E_2/E_1 = 10$ in the material M1. Then the slow increase is present even though its values are not too high. This combination, i.e. $E_2/E_1 = 10$ and displacement loading, produces the lowest stress intensity factor of all the configurations studied independently of the value of gradient index g.

The previous analyses showed the possibility to study the FGMs by the method presented. The influence of the material properties distribution is significant and should be taken into account in the design of FGM structures. A suitable configuration is able to prolong the residual lifetime of the cracked structure and to assist in safe service.

4.3. Influence of the thickness of the FGM layer

A parameter which is often considered during the design of an FGM structure is the thickness of the FGM layer. For the thickness of the FGM, $t_{FGM} = 0$, the step change of material properties is assumed and this interface corresponds to the connection of two materials which do not allowed any diffusion. A connection of this type produces significant shear stresses, both positive and negative. The greater the difference in material properties, the greater the step in shear stress. In order to minimize these shear stresses the thickness of the FGM layer should be $t_{FGM} > 0$. The analysis of the influence of the t_{FGM} has been performed to show the effect on the stress intensity factor. The geometry studied is the same as in previous analyses, see Fig. 2. The thickness of the FGM layer t_{FGM} varied from 0 to 26 mm. In case of $t_{FGM} = 0$ the bi-material interface is modelled. The total thickness of the specimen is $w = 30$ mm. The linear change of the Young's modulus had been assumed. The results of the simulations are shown in Fig. 10. Even though the thicker FGM layer causes a decrease in the shear stresses it negatively affects the fracture behaviors of the structure. However, the positive effect of the decrease in shear stress ultimately proves to be an advantage that more significantly contributes to the longer lifetime of the structure than a low stress intensity factor by itself.

5. Conclusions

The aim of this paper was to study complex FGM structures from a fracture mechanics point of view. The idea of the discretization of the material properties is proposed here. The advantage of this principle is the fine FE mesh in locations where the gradient of the material properties is higher. The sensitivity analysis of the step factor s has been carried out. It was found that for $s = 1.05$ and $s = 1.1$ the results are identical. A good accuracy was also found for $s = 1.2$. The use of $s = 1.1$ for ratios $E_2/E_1 > 1$ and $s = 1.2$ for ratios $E_2/E_1 < 1$ can be recommended. The analyses of the influence of the material properties distribution were performed on the basis of this discretization principle. Three types of shapes were studied – linear change, prompt change from material M1 to material M2, and strong change closer M2, see Fig. 7.

The values of the stress intensity factor were calculated for boundary conditions corresponding to displacement loading. In general, the material properties distributions where the strong change closer to M2 was assumed showed a radical increase in the stress intensity factor in the vicinity of the interface between FGM layer and material M2. This effect positively influence the lifetime of the structure.

The last analysis was focused on simulations of various thicknesses of the FGM layers. Even though the thicker FGM layer causes a decrease in shear stress, it negatively affects the fracture behaviours of the structure. The analyses proved that with an increase in t_{FGM} the stress intensity factor is even greater. However, the ultimate positive effect of a decrease in shear stress is an advantage that contributes to the longer lifetime of the structure more significantly than a low stress intensity factor on its own.

The principle proposed here can be useful for the FE simulation of 3D FGM structures to minimize computational time and for effective simulations. The results can also help in the design process of FGM structures as well as in the educational field.

Acknowledgements

This research was supported by grants 101/09/J027 and 106/09/H035 of the Czech Science Foundation.

References

[1] Anderson, T. L., *Fracture Mechanics - Fundamentals and Applications*, CRC Press Inc., 1995.
[2] Bahr, H.-A., et al., Cracks in functionally graded materials. Material Science and Engineering A 362, 2–16, 2003.
[3] Bao, G., Wang, L., Multiple cracking in functionally graded ceramic/metal coatings. Int. J. Solids and Structures 32, 2 583–2 871, 1995.
[4] Delale, F., Erdogan, F., The crack problem for a nonhomogeneous plane. Journal of Applied Mechanics 50 (3), 609–614, 1983.
[5] Delale, F., Erdogan, F., On the mechanical modeling of the interface region in bonded half-planes. Journal of Applied Mechanics 55, 317–324, 1988.
[6] Dolbow, J. E., Nadeau, J. C., On the use of effective properties for the fracture analysis of microstructured materials. Engineering Fracture Mechanics 69, 1 607–1 634, 2002.

[7] Dolbow, J. E., Gosz, M., On the computation of mixed-mode stress intensity factors in functionally graded materials. International Journal of Solids and Structures 39, vol. 9, 2 557–2 574, 2002.

[8] Erdogan, F., Fracture mechanics of functionally graded materials. Final Technical Report no. 533033. Prepared for U.S. Air Force Office of Scientific Research. 1996.

[9] Erdogan, F., Wu, B. H., The surface crack problem for a plate with functionally graded properties. Journal of Applied Mechanics 64, 449–456, 1997.

[10] Chi, S.-h., Chung, Y.-L., Cracking in coating – substrate composite with multilayered and FGM coatings. Engineering Fracture Mechanics 70, 1 227–1243, 2003.

[11] Cho, J. R., Ha, D. Y., Volume fraction optimization for minimizing thermal stress in Ni-Al2O3 functionally graded materials. Material Science and Engineering A334, 147–155, 2002.

[12] Long, X., Delale, F., The mixed mode crack problem in an FGM layer bonded to a homogenous half-plane. Int. J. Solids and Structures 42, 3 897–3 917, 2005.

[13] Ševčík, M., Hutař, P., Náhlík, L., Knésl, Z., Graded strip model of the polymer weld. Proceeding of International Conference on Computational Modelling and Advanced Simulations, Bratislava, Slovak Republic, 30 June–3 July 2009.

[14] Tilbrook, M. T. et al., Crack propagation in graded composites. Composite Science and Technology 65, 201–220, 2005.

[15] Tilbrook, M. T. et al., Crack propagation paths in layered, graded composites. Composites B 37, 490–498, 2006.

[16] Untao, S., Amada, S., Fracture properties of bamboo. Composites: Part B 32, 451–459, 2001.

[17] Wang, B. L., Mai, Y.-W., Noda, N., Fracture mechanics analysis model for functionally graded materials with arbitrarily distributed properties. International Journal of Fracture 116, 161–177, 2002.

4

Mathematical modeling of a biogenous filter cake and identification of oilseed material parameters

J. Očenášek[a,*], J. Voldřich[a]

[a]*New Technologies Research Centre, University of West Bohemia, Univerzitní 8, 306 14 Plzeň, Czech Republic*

Abstract

Mathematical modeling of the filtration and extrusion process inside a linear compression chamber has gained a lot of attention during several past decades. This subject was originally related to mechanical and hydraulic properties of soils (in particular work of Terzaghi) and later was this approach adopted for the modeling of various technological processes in the chemical industry (work of Shirato). Developed mathematical models of continuum mechanics of porous materials with interstitial fluid were then applied also to the problem of an oilseed expression. In this case, various simplifications and partial linearizations are introduced in models for the reason of an analytical or numerical solubility; or it is not possible to generalize the model formulation into the fully 3D problem of an oil expression extrusion with a complex geometry such as it has a screw press extruder.

We proposed a modified model for the oil seeds expression process in a linear compression chamber. The model accounts for the rheological properties of the deformable solid matrix of compressed seed, where the permeability of the porous solid is described by the Darcy's law. A methodology of the experimental work necessary for a material parameters identification is presented together with numerical simulation examples.

Keywords: filter cake, fractionation process, permeability, continuum mechanics

1. Introduction

Oilseed expression has become an important technology not only in the food industry, but also for biofuel manufacturing with rapeseed oil as the most common component. The oilseed expression process could be undertaken in linear compression chambers or in screw press extruders. Nowadays almost the whole rapeseed oil production and the production of other vegetable oils is processed by extruders. This led to an extensive development of experimental programs with the aim of the farther screw extruders optimization (see e.g. [5, 6]). On the other hand, no convenient mathematical models of the oilseed expression were proposed for the screw press geometry, or there are no computational software tools necessary for the numerical modeling applying existing models [11]. An exception is the work of [9] based on experiences of Japanese chemical engineers in comparable technologies of fractional processes. Unfortunately the proposed methodology is strongly based on the engineering intuition and a number of empirical relations that limit its generalization.

This situation is much different for the oilseed expression in a linear compression chamber. In this case, extensive experimental programs of the expression process were carried out [10, 6, 12, 5, 3], as well as their appropriate mathematical modeling were performed for the identification of relevant material parameters of various oilseeds.

---

*Corresponding author. e-mail: ocenasek@ntc.zcu.cz.

The work of Shirato [7] could be considered as one of the most evolved analytical approach to model the compression of a filter cake. His model is a limiting case of the conservation laws model, as a number of assumptions need to be made in order to be able to derive an analytical solution. Nevertheless, the conservation laws 1D models based on mass and momentum balances were solved in the last decade (see e.g. [3]). Unfortunately, these models were formulated in such a manner that their three-dimensional generalization for a screw press geometry is not feasible.

Nomenclature

0	suffix related to the initial state
f	suffix related to the fluid
s	suffix related to the solid matrix
c	empirical parameter of the equation suggested by Tiller and Yeh [–]
K	permeability [m^2]
p	liquid pressure [Pa]
p_s	solid compressive pressure [Pa]
$p_f = \Phi p$	effective liquid pressure [Pa]
P	externally applied pressure of the volume V [Pa]
q	superficial liquid flow [m/s]
S	piston area [m^2]
t	time [s]
v	solid phase velocity [m/s]
V	local volume [m^3]
V_0	the volume at $t = 0$ [m^3]
V_s	local volume of the solid phase [m^3]
V_f	local volume of the liquid phase [m^3]
w	liquid velocity relative to the solid velocity [m/s]
x	spatial coordinate (distance from the filter) [m]
α	specific cake resistance [m/kg]
γ	material parameter of the solid matrix analogous to the Biot coefficient [–]
ϵ	(natural) volume strain of the solid matrix [–]
η	viscosity of the dash pot [Pa · s]
μ	liquid viscosity [Pa · s]
$\Phi = V_f/V$	porosity [–]
$\Phi_0 = V_{f0}/V_0$	the porosity at $t = 0$ [–]
ρ_s	(intrinsic) solid phase density [kg/m^2]
ω	spatial convective (material) coordinate [m]

$$d\epsilon = dV/V \quad \epsilon = \ln\left(\frac{V}{V_0}\right) \quad V = V_s + V_f \quad P = p_s + p_f = p_s + \Phi p \quad q = \Phi w$$

During the oilseed compression process the bed of seeds becomes more compacted due to the expulsion of air. After a short time, when the pressure raises the so called oil point, the seeds are crushed, whereby the seed structure containing the oil is damage and the oil is freed. This allows the oil to drain through the bed of crushed nibs and allows it to fraction. During this stage the solid matrix and the freed oil carry the external load. Therefore, the biogenous filter cake can be, within the mathematical model, considered as a two-phase medium that consists of a porous material (the solid matrix) saturated with an interstitial fluid (the vegetable oil) as the second phase. The theory of the filtration and consolidation of saturated porous media had received a lot of attention by geomechanics, ground engineering and engineering of building materials. The deformation behavior of the porous media are, within these fields, most frequently described by the Biot constitutive model (see e.g [1]). Unfortunately, the Biot model does not satisfactorily cover the case of large volume deformations, which are generally not of a big interest for the area of ground engineering or geomechanics, where the large volume deformations are undesirable or related to a failure state.

This paper is arranged as follows. In the following section of the paper a mathematical model of the filter cake compression process is derived from the basic conservation laws, where the primary unknown quantities are the (natural) volume strain ϵ of the solid matrix, the effective oil pressure p_f and the solid velocity v as functions of the spatial coordinate x and the time t. The Darcy's law is used to describe the local cake permeability and the viscoelasticity of the solid matrix is considered. The velocity v (following the Euler continuum approach) is selected as a primary unknown quantity, since it is naturally possible to extend the model to the three-dimensional case of the screw press with a fixed boundary, as it was presented in [11]. In the case of a linear compression chamber there is a moving piston, meaning that the problem with a moving boundary is obtained, though this difficulty can be overcome by applying the spatial convective coordinate. In the section 3 setups of various experiments are proposed, which should be sufficient for the identification of material parameters applied in the mathematical model. The section 4 is devoted to identification of material parameters. Related numerical simulations based on the presented model were performed and the comparison of the simulation results with experimental observations published in the literature is given. Finally, the section 5 provides discussion and summary of achieved results.

2. Mathematical model for fractionation mechanism

The compression process of a filtr cake, as shown in Fig. 1, is considered in this section. With regard to the large volume changes of the filter cake, the volume changes caused by the compressibility of the liquid and the intrinsic compressibility of the solid phase can be neglected. Thus, in following the solid phase and the liquid are assumed as incompressible with a constant density ρ_s and ρ_f, respectively.

Additionally it is assumed that the expression process is one-dimensional (therefore the friction between the compression cake and the walls of the compression chamber is negligible),

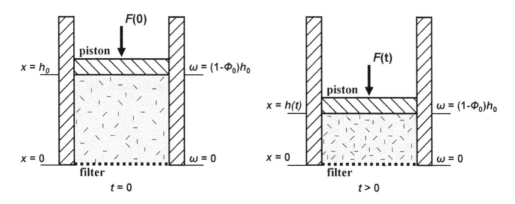

Fig. 1. The expression process at the start and for the time $t > 0$ with the applied force $F(t)$

that no solids pass trough the filter medium, that the effect of inertia and gravity is negligible, that the resistance of the filter medium is negligible compared to that of the filter cake and that the local porosity Φ is a smooth function of both spatial position and time.

2.1. Mass balance

The solid mass of a volume V_0 of the solid matrix is $m_s = \rho_s V_{s0} = \rho_s(V_0 - V_{f0}) = \rho_s V_0(1 - \Phi_0)$. At the time t the corresponding part of the solid matrix occupies a volume V, and by analogy we obtain $m_s = \rho_s V(1 - \Phi)$, thus $1 - \Phi = (1 - \Phi_0)V_0/V$. Since $-\epsilon = \ln(V_0/V)$, the following relation between the porosity Φ and the volume strain ϵ can be derived

$$\Phi = 1 - (1 - \Phi_0)\exp(-\epsilon). \tag{1}$$

Let ρ be the "distributed" density of the solid matrix in a volume V, i.e. $\rho V = \rho_s V_s$. From the solid mass balance it holds that $\rho V = \text{const.}$ and so $dV/V = -d\rho/\rho$. As $d\epsilon = dV/V$, we obtain $d\epsilon = -d\rho/\rho$ and furthermore $\rho\dfrac{\partial \epsilon}{\partial t} = -\dfrac{\partial \rho}{\partial t}$, $\rho\dfrac{\partial \epsilon}{\partial x} = -\dfrac{\partial \rho}{\partial x}$. Applying the last two relations to the fundamental equation of the mass balance $\dfrac{\partial \rho}{\partial t} + \dfrac{\partial}{\partial x}(\rho v) = 0$, we find

$$\frac{\partial \epsilon}{\partial t} + v\frac{\partial \epsilon}{\partial x} - \frac{\partial v}{\partial x} = 0. \tag{2}$$

Since we assume that the densities ρ_s and ρ_f are constant, we can also write fundamental equations of the mass balance in the form

$$\frac{\partial \Phi}{\partial t} + \frac{\partial}{\partial x}(\Phi(w + v)) = 0, \qquad \frac{\partial(1 - \Phi)}{\partial t} + \frac{\partial}{\partial x}((1 - \Phi)v) = 0.$$

Now, we obtain by their summation $\dfrac{\partial v}{\partial x} + \dfrac{\partial}{\partial x}(\Phi w) = 0$, i.e.

$$\frac{\partial v}{\partial x} + \frac{\partial q}{\partial x} = 0 \tag{3}$$

where $q = \Phi w$ is the superficial flow of the oil.

2.2. Force balance

We can assume that shear stress is zero. The one-dimension equation of the force balance is then

$$-\frac{\partial P}{\partial x} = 0, \tag{4}$$

where P is the external pressure acting on a local volume V. Consequently, P is the function of time t only, $P(t) = F(t)/S$, where S is the area of the piston, and

$$P(t) = p_s(x, t) + p_f(x, t). \tag{5}$$

2.3. Constitutive equations

2.3.1. Cake permeability

The relationship between the liquid velocity w and the liquid pressure gradient p_f is determined by the local cake permeability. The (intrinsic) liquid flow within pores of the solid matrix can be assumed to be laminar due to the Reynolds number (see e.g. [3]) and therefore the relationship can be described by the Darcy's law

$$q = -\frac{K}{\mu} \frac{\partial p_f}{\partial x}, \tag{6}$$

where K is the permeability and μ is the dynamic viscosity of the oil. In the case of fluid mechanics in porous mediums it is standard practice to use the Kozeny-Carman's empirical law to evaluate the permeability of the medium in function of its porosity Φ. However, the Tiller-Yeh's approach (see [3, 10]) seems to be more promising when considering cake filtration. If

$$\alpha = \frac{1}{(1 - \Phi)K\rho_s} \tag{7}$$

is now the specific cake resistance, then we obtain by the Tiller-Yeh's empirical law

$$\alpha = \alpha_0 \left(\frac{1 - \Phi}{1 - \Phi_0}\right)^c, \quad \text{i.e.} \quad K(\Phi) = \frac{(1 - \Phi_0)^c}{\alpha_0 \rho_s (1 - \Phi)^{c+1}}. \tag{8}$$

2.3.2. Solid matrix compressibility

Experimental data published in the literature indicate that the solid matrix deformation behavior is not only elastic, but also exhibits a creep character. Therefore, modeling of the oil expression solely by recourse to the theory of poro-elasticity would be insufficient. There are several empirical relations published in the literature (see e.g. [3]). Within this paper, we chose to adopt the theory of Voigt in the generalized form

$$\mathcal{F}(\epsilon) + \eta \frac{d\epsilon}{dt} = -p_s + \gamma p, \tag{9}$$

where the $\mathcal{F}(\epsilon)$ has the meaning of the pressure response of the solid matrix subjected to a volume deformation ϵ, in the case of the zero fluid pressure (for the linear case $\mathcal{F}(\epsilon) = E\,\epsilon$). The parameter $\eta \neq 0$ is the viscosity of the dash pot and γ is a parameter analogous to the Biot coefficient taking into account the influence of the oil pressure.

2.4. Summary

When folowing functions $\epsilon = \epsilon(x, t)$, $p_f = p_f(x, t)$, $v = v(x, t)$ are taken as primary unknows, the mathematical model of the fractionation mechanism can be formulated as

$$\frac{d\epsilon}{dt} = \frac{\partial v}{\partial x}, \tag{10a}$$

$$\frac{\partial}{\partial x}\left(\eta \frac{\partial v}{\partial x}\right) + \frac{\partial}{\partial x}\left(\mathcal{F}(\epsilon) - (1 + \frac{\gamma}{\Phi(\epsilon)})p_f\right) = 0, \tag{10b}$$

$$\frac{\partial}{\partial x}\left(\frac{K(\Phi(\epsilon))}{\mu} \frac{\partial p_f}{\partial x}\right) - \frac{\partial v}{\partial x} = 0 \tag{10c}$$

with the relation (1) for $\Phi = \Phi(\epsilon)$ and the relation (8) for $K = K(\Phi)$. The presented approach is especially convenient for its generalization to the 3D problem of the oil expression in a screw press extruder. Here applied material parameters are μ, α_0, c, η, γ, the function $\mathcal{F} = \mathcal{F}(\epsilon)$ and the initial porosity Φ_0.

The first of the equations (10a) is parallel to the equation (2), because the derivative $\mathrm{d}/\mathrm{d}t$ denotes the material derivative here. The second equation (10b) can be derived by substituting for the term $\mathrm{d}\epsilon/\mathrm{d}t$ from the first equation (10a) into the equation (9) and also for the term p_s from the relation (5), after that by taking the derivative with respect to x and applying the equation (4). Finally, the term (10c) is obtained combining the Darcy's law (6) and the balance equation (3).

In our one-dimensional case an even more simple form can be derived. The external pressure P acting on the (local) volume V does not depend on the spatial variable x and as we know, $P(t) = F(t)/S$, thus the pressure P can be kept in equations. This, at the same time, allows to eliminate the derivative $\partial v/\partial x$ from the related terms as follows. The first equation of the new system can be derived from the equation (9) by substitution for the term p_s from (5) and applying the relation $p = p_f/\Phi$ with respect to different liquid pressure definitions. The second equation is then obtained from the equation (10c) with the help of (10a) by replacing the derivative $\partial v/\partial x$ by the derivative $\mathrm{d}\epsilon/\mathrm{d}t$ which is finally expressed by (11a). For only two unknown quantities $\epsilon = \epsilon(x,t)$ and $p_f = p_f(x,t)$ we now obtain

$$\eta \frac{\mathrm{d}\epsilon}{\mathrm{d}t} + \mathcal{F}(\epsilon) - \frac{\gamma}{\Phi(\epsilon)} p_f = -P(t) + p_f \, , \tag{11a}$$

$$-\frac{\partial}{\partial x}\left(\frac{K(\Phi(\epsilon))}{\mu}\frac{\partial p_f}{\partial x}\right) + \frac{1 + \gamma/\Phi(\epsilon)}{\eta} p_f = (P(t) + \mathcal{F}(\epsilon))\,/\eta \, . \tag{11b}$$

Initial and boundary conditions will be discussed later.

2.5. Formulation using the convective (material) coordinate

With respect to the piston displacement in the linear compression chamber, the part of the domain boundary, where equations (11) are being solved, is moving. To reformulate this problem into a problem with a fixed domain boundary, we will introduce the convective (material) coordinate ω as

$$w(t, x) = \int_0^x (1 - \Phi(t, X))\,\mathrm{d}X \, . \tag{12}$$

It is important to realize that the coordinate $\omega = \omega(t, x(t, A))$ is time independent for a selected material particle A of the solid matrix, i.e. $\omega = \omega(A)$. For the selection and $x = x(t, A)$ it comes out that the integral on the right hand side of (12) is time independent, because by multiplying it with the (constant) density ρ_s and with the piston cross-section area S we obtain the mass of the solid matrix between the filter and a cross-section (perpendicular to the filter cake axis) passing trough the material particle A. It is $\omega(t, 0) = 0$ at the point of the filter and $\omega(t, h(t)) = \omega(0, h_0) = \int_0^{h_0}(1 - \Phi_0)\,\mathrm{d}X = (1 - \Phi_0)h_0$ just under the piston. By virtue of

$$\frac{\partial}{\partial x} = (1 - \Phi)\frac{\partial}{\partial \omega} \quad \text{and} \quad \frac{\mathrm{d}\epsilon(t, x)}{\mathrm{d}t} = \frac{\partial \epsilon(t, \omega)}{\partial t}$$

the problem (11) can be rewritten for the convective coordinate ω. For the unknown functions $\epsilon = \epsilon(t, \omega)$ and $p_f = p_f(t, \omega)$, $0 \leqq \omega \leqq (1 - \Phi_0)h_0$, we then have the system of partial differential equations

$$\eta \frac{\partial \epsilon}{\partial t} + \mathcal{F}(\epsilon) - \frac{\gamma}{\Phi(\epsilon)} p_f = -P(t) + p_f \,, \tag{13a}$$

$$-\frac{\partial}{\partial \omega} \left(\frac{K(\Phi(\epsilon))}{\mu} (1 - \Phi(\epsilon)) \frac{\partial p_f}{\partial \omega} \right) + \frac{1 + \gamma/\Phi(\epsilon)}{\eta(1 - \Phi(\epsilon))} p_f = \frac{1}{\eta} \frac{P(t) + \mathcal{F}(\epsilon)}{1 - \Phi(\epsilon)} \tag{13b}$$

with boundary conditions

$$p_f(t, 0) = 0 \,, \quad \frac{\partial p_f}{\partial \omega}(t, (1 - \Phi_0)h_0) = 0 \tag{14}$$

for the situation at the Fig. 1 and with the initial condition

$$\epsilon(0, \omega) \equiv 0 \,.$$

3. Experimental procedure

Material parameters like the density ρ_s, the viscosity μ and the initial porosity Φ_0 can be estimated in a standard way and reliable values can be found in the literature [12, 2]. Therefore we focus our attention to other material parameters named in the section 2.4, that are c, α_0, η, γ and $\mathcal{F}(\epsilon)$.

3.1. Drained experiment and "steady state" component of the volume deformation

The scheme of a drained experiment is illustrated at the Fig. 1, where the acting force $F(t)$ rises and settles at the constant value of F_∞. Farther we put our attention only to the initial unloaded configuration and the final steady state. The initial state is defined by the values h_0, $P_0 = F(0)/S = 0$, i.e. $p_{f0} = p_{s0} = 0$. The final steady state is reached, when no more oil is being expeled from the filter cake and the height of the cake remains unchanged, so the gradients of quantities p_f and ϵ are zero.

Then, at the time t_∞, when the steady state is reached precisely enough, it is $h(t_\infty) = h_\infty$, $p(t_\infty, x) = 0$, $\mathrm{d}\epsilon/\mathrm{d}t(t_\infty, x) = 0$, $\epsilon(t_\infty, x) = \mathrm{const.} = \epsilon_\infty$ and $v(t_\infty, x) = 0$ for all $x \in (0, h_\infty)$. This yields $p_s(t_\infty, x) = P(t_\infty, x) = P_\infty = F(t_\infty)/S$ and it is $\mathcal{F}(\epsilon_\infty) = -P_\infty$ for the corresponding deformation value $\epsilon_\infty = \ln(h_\infty/h_0)$. The relation $\epsilon \to \mathcal{F}(\epsilon)$ can be identified from various steady state configurations, characterized by different values of h_∞.

3.2. Undrained experiment and the material parameter γ

Within this subsection we will discuss two different steady state configurations of the undrained experiment as illustrated at the Fig. 2.

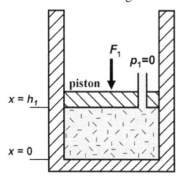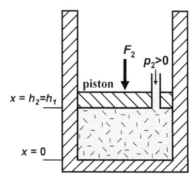

Fig. 2. Two steady states of an undrained compressive test for the identification of the parameter γ

In both cases, the deformation ϵ and the intrinsicl fluid pressure p are constant in respect to the x. The experiment is performed so that the height of the filter cake is the same in both cases, i.e. $h_2 = h_1$, and so $\epsilon_2 = \epsilon_1 = \ln(h_1/h_0)$. However $p_1 = 0$ in the first case, while the invariant pressure $p_2 > 0$ of the oil is kept under the piston in the second case. Perceiving that it is $d\epsilon/dt = 0$ and $p_s = P - \Phi p = F/S - \Phi p$ for steady state, then by applying relations (9) and (1), it yields to the result

$$\gamma = \frac{1}{p_2}(F_1/S - F_2/S - \Phi p_2) \quad \text{with} \quad \Phi = 1 - (1 - \Phi_0)\frac{h_0}{h_1} \,.$$

These experiments for different values of ϵ_1, Φ_0 and p_2 reveal, how effectively the parameter γ could be considered as a constant and how the dependence on these quantities is significant.

3.3. Drained experiment and permeability

The scheme of a drained experiment, supplied by an oil pressure line leading under the piston, is presented at the Fig. 3. Let us consider a steady state, when the piston stops to shift and when the pressure field as weel as the total oil flow q_1 are stabilized. Then $d\epsilon/dt(t_\infty, x) = 0$ and $v(t_\infty, x) = 0$ for all $x \in (0, h_1)$. With regard to the present oil flow and non-zero pressure gradient p, the deformation ϵ and the porosity Φ are not constant. By virtue of (11a) it can be written that

$$p_f(x) = \{\mathcal{F}(\epsilon(x)) + F_1/S\}/\{1 + \gamma/\Phi(\epsilon(x))\} \,,$$

which by substitution into (11b) leads to a differential equation for the deformation $\epsilon(x)$

$$\frac{\partial}{\partial x}\left(\frac{K(\Phi(\epsilon(x)))}{\mu}\frac{\partial}{\partial x}\left(\frac{\mathcal{F}(\epsilon(x)) + F_1/S}{1 + \gamma/\Phi(\epsilon(x))}\right)\right) = 0 \,,$$

where functions $\Phi = \Phi(\epsilon)$ and $K = K(\Phi)$ have the form of (1) and (8), respectively.

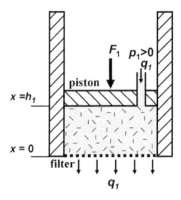

Fig. 3. Steady state of the drained compressive test with oil flow for the identification of the permeability

The corresponding boundary condition for $\epsilon = \epsilon(0)$ at the side of the filter can be derived by resolving the following algebraic equation

$$0 = (\mathcal{F}(\epsilon) + F_1/S)/(1 + \gamma/\Phi(\epsilon)) \,,$$

and by analogy, the boundary condition for $\epsilon = \epsilon(h_1)$ at the side of the piston, by solving

$$\Phi(\epsilon)\,p_1 = (\mathcal{F}(\epsilon) + F_1/S)/(1 + \gamma/\Phi(\epsilon)) \,.$$

Referred material parameters α_0 and c of the Tiller-Yeh's empirical law (8), which characterize the permeability K, have to be fitted in such a way that the calculated oil flow rate matches well the real flow rate q_1 from the experiment, i.e. that the following relation is fulfilled as precisely as possible for all $x \in (0, h_1)$

$$\frac{K(\Phi(\epsilon(x)))}{\mu} \frac{\partial p_f}{\partial x}(x) = q_1 .$$

3.4. *Drained experiment and the viscosity of the solid matrix*

In the case of identifying the solid matrix viscosity η, the analysis of the steady state configuration in the experiment from the Fig. 1 is insufficient and a complete record of the continuous process of the filter cake compression is necessary. The fitting of the material parameter η has to be then done so that the compression process calculated by means of equations (13a) and (13b) corresponds well to the experimental data.

4. Transient modes and identification of material parameters

The piecewise linear Galerkin method and the explicit integration scheme was applied for numerical simulation of the unsteady problem defined by equations (13a) and (13b).

4.1. *Material parameters fitting*

As mentioned, material parameters μ, ρ_s were taken from the literature [12] and [2] and the parameter γ was taken to be zero for simplicity, because results corresponding to the test from Fig. 2 were not published. Remaining parameters η, c, α_0 and the relation $\mathcal{F}(\epsilon)$ were determined by fitting of our numerical results of the problem (13)–(14) to data from two different type of experiments, both adopted from the literature [12].

The first experiment is performed in the compression chamber with a setup corresponding to the scheme at Fig. 1 and continuously records the oil yield U_t as a function of time, when a constant pressure of 30 MPa is applied. The yield U_t is a mass fraction of the expeled oil at the time t to the initial oil content and thus it is defined as

$$U_t = \frac{h_0 - h(t)}{\Phi_0 h_0} , \tag{15}$$

The second type of experiment determines the total oil yield U_{10min} for various pressure values P applied by the piston in a linear compression chamber of the same configuration (see Fig. 4), where the period of 10 minutes was estimated in the work of Willems [12] as sufficient to characterize the expression process. However, there are no published experiments confirming that an equilibrium (i.e. a steady state) is reached after this period, especially for high values of the pressure P. Therefore the $\mathcal{F}(\epsilon)$ can not be identified separately from parameters c and α_0 with enough accuracy as proposed in the section 3.1, because the assumption of the steady state is not sufficiently fulfilled.

We assume that the pressure $\mathcal{F}(\epsilon)$ will increase with the growing deformation following the relation

$$\mathcal{F}(\epsilon) = A \frac{\epsilon}{\bar{\epsilon} - \epsilon} , \tag{16}$$

where $\bar{\epsilon}$ is the volume deformation at the state with the zero oil content $\Phi = 0$, i.e. $\bar{\epsilon} = \ln(1 - \Phi_0)$ and the parameter A is to be determined by fit to experimental data.

The discrepancy at low pressure values, especially for $P = 10$ MPa is most probably caused by a threshold pressure called oil point, which indicates pressure at which the oil emerges from a seed kernel during mechanical oilseed expression. The typical value of the oil point for rapeseed is approximately 7–8 MPa. This threshold pressure is not yet incorporated in our model, so the oil yield at low pressure values is notably overestimated. Nevertheless, our research focuses to design extruders able to reach high yields so the filter cake behavior at high pressure states gains more interest.

Parameters A, η, c and α_0 were then fitted to correspond to the experimental data, from Figs. 4 and 5. Summary of the parameter values is given in Tab. 1.

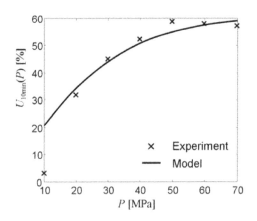
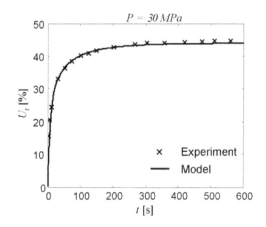

Fig. 4. The oil yield U_{10min} for various values of the external pressure P

Fig. 5. Time dependence of the oil yield U_t for $P = 30$ MPa

Table 1. Material parameters of rapeseed

A	ρ_s	μ	Φ_0	η	α_0	γ	c
[MPa]	[kg/m^3]	[Pa · s]	[–]	[GPa]	[m/kg]	[–]	[–]
52	1 052	0.055	0.467	1.8	1.1×10^{11}	0	12

It has to be stressed that additional experiments (conformable with Figs. 2, 3) added to published ones in [12] should be performed, because otherwise more than one parameter set can be found to aproximate well the experimental data mentioned. For instance, after setting the form of $\mathcal{F}(\epsilon)$ to be of (16), the viscosity η can be selected from quite wide range, while the parameter c and α_0 can be still adjusted to match a good fit of experiments.

4.2. Modeling result analysis

By analyzing the simulation results we can observe that the permeability of the filter cake decreases for higher values of exponent c to the extent that the steady state is not reached within the period of 10 minutes and the time necessary to approach a steady state can be in order of magnitudes longer. For the parameter set from Tab. 1 this effect can be seen for higher loads $P = 60$ and 70 MPa, how we can deduce for example from Figs. 6, 7 and 8.

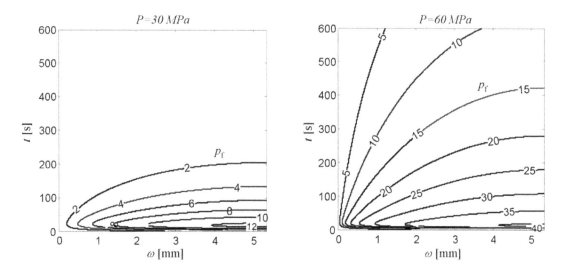

Fig. 6. Contour of the fluid pressure p_f, applied external pressure (a) $P = 30$ MPa and (b) $P = 60$ MPa

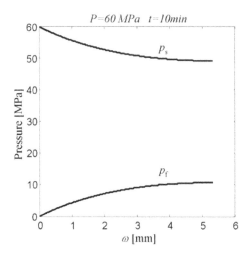

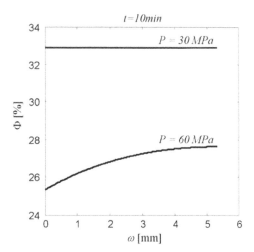

Fig. 7. Distribution of pressures p_f and p_s through the filter cake for $t = 10$ min and $P = 60$ MPa

Fig. 8. Distribution of the porosity Φ through the filter cake for $t = 10$ min and for external pressures $P = 30$ MPa and $P = 60$ MPa

5. Conclusions and discussion

In the chapter 2 we have proposed the mathematical description of oil seed materials, which is covering their rheological properties as well as the permeability of these porous materials by means of the Darcy's law, where the appropriate description of the permeability turns out to be the Tiller-Yeh's empirical law. The proposed mathematical model also covers the possible effect of oil-pressure on the solid matrix behavior trough a parameter analogous to the Biot coefficient.

The third chapter presents three different experiments in a linear compression chamber whose results are sufficient for the identification of suggested model parameters. Unfortunately, experimental results published up to now (e.g. for rapeseed see [12, 2]) are insufficient and more or less pertain only to the test from Fig. 1. Therefore, during the filter cake compression process, it is difficult to distinguish the deformation part rising from the rheological behavior of the solid matrix from effects related just to the permeability.

Thus we can find markedly different material parameters in the literature (e.g. compare [12] and [3] for rapeseed parameters) for the commonly applied Shirato model [7, 5]. Hence, the results of the parameter identification that we present in chapter 4 are only illustrative.

Insufficient experimental results also pose fundamental questions of the filter cake behavior at higher external pressures. Since from published experiments it is not possible to reveal, how long is the relaxation time to reach the steady state or even, whether eventually the pores of the solid matrix get completely closed, i.e. achieving practically zero permeability at low porosity. In such case the Tiller-Yeh's empirical law would have to be modified.

Finally, it has to be remarked that the height of the filter cake is assumed to be sufficiently small in comparison to the diameter of the compression chamber, so the friction of the filter cake with the chamber wall can be neglected and that the temperature field of the filter cake is approximately uniform.

Acknowledgements

The paper is based on work supported by the Ministry of Industry and Trade of the Czech Republic under the research project FR-TI 1/369. This project is solved in collaboration with FARMET, a. s., Česká skalice.

References

[1] Detournay, E., Cheng, A., Fundamentals of poroelasticity, Pergamon Press, 1993.

[2] Faborode, M. O., Favier, J. F., Identification and significance of the oil-point in seed-oil expression, J. Agric. Engng Res., 65 (1996), p. 335–345.

[3] Kamst, G. F., Filtration and expression of palm oil slurries as a part of the dry fractionation process, Dissertation, Technische Universiteit Delft, 1995.

[4] Olivier, J., Vaxelaire, J., Vorobiev, E., Modelling of cake filtration: An overview, Separation Science and Technology 42 (2007), p. 1 667–1 700.

[5] Raß, M., Zur Rheologie des Biogenen Feststoffs unter Kompression am Beispiel Geschäälter Rapssaat, Dissertation, Universität Gesamthochschule Essen, 2001.

[6] Schein, Ch., Zum Kontinuierlichen Trennpressen Biogener Feststoffe in Schneckengeometrien am Beispiel Geschälter Rapssaat, Dissertation, Universität Duisburg-Essen, 2003.

[7] Shirato, M., Murase, T., Iwata, M., Hayashi, N., Deliquoring by expression – Part 1 Constant-pressure expression, Drying Technology 4 (3) (1986), p. 363–386.

[8] Shirato, M., Murase, T., Iwata, M., Pressure profile in a power-law fluid in constant-pitch, straight-taper and decreasing pitch screw extruders, International Chemical Engineering, 23 (2) (1983), p. 323–331.

[9] Vadke, V. S., Sosulski, F. W., Shook, C. A., Mathematical simulation of an oilseed press, JAOCS 65 (10) (1988), p. 1 610–1 616.

[10] Venter, M. J., Kuipers, N. J. M., de Haan, A. B., Modelling and experimental evaluation of high-pressure expression of cocoa nibs, Journal of Food Engineering 80 (2007), p. 1 157–1 170.

[11] Voldřich, J., Mathematical modelling of the dry fractionation process of oilseed in presses with the linear or screw geometry (in Czech), Research Report, New Technology Research Center, UWB, 2009.

[12] Willems, P., Gas assisted mechanical expression of oilseeds, Dissertation, Universiteit Twente, Nederland, 2007.

5

Analysis of composite car bumper reinforcement

V. Kleisner[a,*], R. Zemčík[a]

[a]*Faculty of Applied Sciences, University of West Bohemia, Univerzitní 22, 306 14 Plzeň, Czech Republic*

Abstract

The presented work summarizes the present state of car passive safety testing according to European methodologies. The main objective is to analysis a bumper reinforcement made of composite materials. The bumper is tested according to RCAR (Research Council for Automobile Repairs) methodology using numerical simulation. Individual proposed variants are compared with the existing steel construction which does not comply with manufacturers specifications. The PAM-Crash software is used for the simulation. The numerical model is using shell elements and the Ladevèze model for the description of behaviour of composite materials. The methodology for the set-up of the numerical model in PAM-crash is firstly validated by comparison of experiment and analytical results.

Keywords: passive safety, finite element method, RCAR, composite materials, bumper reinforcement

1. Introduction

Car accidents are happening every day. Most drivers are convinced that they can avoid such troublesome situations. Nevertheless, we must take into account the statistics – ten thousand dead and hundreds of thousands to million wounded each year. These numbers call for the necessity to improve the safety of automobiles during accidents. As a result most present-day automobiles have at least safety belts with retractors and airbags [6]. However, car accident does not necessarily mean bodily harm. In a low-speed car accident only physical damage occurs, assuming that basic safety regulations, such as fastened seat belts, are kept.

This work focuses on the application of composite materials in a frontal bumper system. In recent years, composite materials have been more frequently used in industry and they have progressively replaced metal materials. This trend is caused by their high strength (and stiffness) to mass ratio and the ability to produce parts with required mechanical properties. The main objective is to analysis the bumper reinforcement made of composite materials and to compare it with the original steel bumper reinforcement in terms of its stiffness and damage behaviour, and mass reduction.

2. European car testing

2.1. EuroNCAP

EuroNCAP [5] company was founded in 1997 and has two main objectives. Firstly, it provides information about comparable automobile safety rating. Secondly, it tries to motivate producers to improve automobile safety and thereby reduces the number of wounded passengers. This

---

*Corresponding author. e-mail: kleisner@kme.zcu.cz.

task is progressively fulfilled in conjunction with automobile producers. The testing stems from EEVC (European Enhanced Vehicle-safety Committee) procedures.

The tests implemented by EuroNCAP are:

- frontal impact,

- side impact,

- pole impact,

- child protection,

- pedestrian protection.

2.2. RCAR

RCAR [4], the Research Council for Automobile Repairs, is an international organization that works towards reducing insurance costs by improving automotive damage ability, repair ability, safety and security by low-speed offset car crash test.

The test vehicle speed shall be 15 km/h within a one-meter distance from the barrier. The barrier offset of the test vehicle is 40 %. The barrier is skewed in a 10 degrees angle and has a radius of 150 mm as shown in Fig. 1.

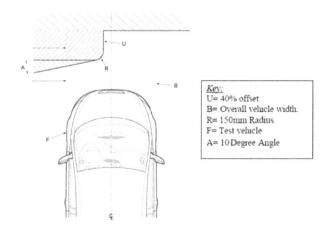

Key:
U= 40% offset.
B= Overall vehicle width.
R= 150mm Radius
F= Test vehicle
A= 10 Degree Angle

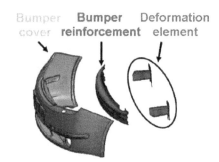

Bumper **Bumper** Deformation cover **reinforcement** element

Fig. 1. Scheme of RCAR test

Fig. 2. Frontal bumper system components

3. Bumpers

Bumpers are fixed on the front and on the back side of a car and serve as its protection. They reduce the effects of collision with other cars and objects due to their large deformation zones. The bumpers are designed and shaped in order to deform itself and absorb the force (kinetic energy) during a collision. The whole frontal bumper system consists the following parts (also seen in Fig. 2):

- the cover,

- the mechanical and deformation energy absorber,

- the bumper reinforcement.

4. Composite damage model in PAM-Crash

The Ladevèze model [3] is dedicated to the numerical simulation of unidirectional continuous fiber reinforced composite materials. Unlike the heterogeneous "biphase" model, the Ladevèze model does not treat the two phases separately (fibers and matrix). Instead the composite ply is described by homogeneous continuum mechanics. In this model some damages related to experimentally observed phenomena are included (see Fig. 3).

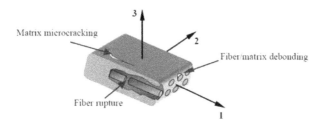

Fig. 3. Composite ply damages

The constitutive relationship of the Ladevèze model can be expressed as follows [3]:

$$
\begin{bmatrix} \varepsilon^e_{11} \\ \varepsilon^e_{22} \\ 2\varepsilon^e_{12} \\ 2\varepsilon^e_{23} \\ 2\varepsilon^e_{13} \end{bmatrix} = \begin{bmatrix} 1/E_1 & -\nu^0_{12}/E_1 & 0 & 0 & 0 \\ -\nu^0_{12}/E_1 & 1/E_2 & 0 & 0 & 0 \\ 0 & 0 & 1/G_{12} & 0 & 0 \\ 0 & 0 & 0 & 1/G_{23} & 0 \\ 0 & 0 & 0 & 0 & 1/G_{12} \end{bmatrix} \begin{bmatrix} \sigma_{11} \\ \sigma_{22} \\ \sigma_{12} \\ \sigma_{23} \\ \sigma_{13} \end{bmatrix},
\tag{1}
$$

where

1. fiber direction: $E_1 = E_1^0(1 - d^{ft})$

2. transverse direction: $E_2 = E_2^0(1 - d')$

3. for shear: $G_{12} = G_{12}^0(1 - d)$

In these equations d^{ft} is the fiber damage constant and d' and d are the matrix damage constants.

Elastic matrix damaging behavior

The matrix related damages are taken into account by two scalar variables, d and d'. These variables express an experimentally displayed phenomena: parameter d quantifies the damage which comes from the debonding between fibers and matrix, whereas parameter d' is related to the damage due to the microcracking of the matrix parallel to the fiber direction.

The damage functions, Z_d and Z'_d, associated respectively with d and d' are defined by the expressions

$$
\frac{\partial E_D}{\partial d} = Z_d = \frac{1}{2} \frac{\sigma^2_{12} + \sigma^2_{13}}{G^0_{12}(1 - d)^2},
\tag{2}
$$

$$
\frac{\partial E_D}{\partial d'} = Z'_d = \frac{1}{2} \frac{\langle \sigma_{22} \rangle^2_+}{E^0_2(1 - d')^2},
\tag{3}
$$

where (2) is shear damage and (3) is transverse damage. E_D is elastic strain energy and $\langle a \rangle_+ = a$, if $a > 0$, else $a = 0$.

The damage evolution functions over time t are defined as

$$Y(t) = \sup_{\tau \geq t} \sqrt{Z_d(\tau) + bZ'_d(\tau)} \tag{4}$$

$$Y'(t) = \sup_{\tau \geq t} \sqrt{Z'_d(\tau)}. \tag{5}$$

The damage values d and d' are calculated from

$$d = \frac{\langle Y(t) - Y_0 \rangle_+}{Y_c}, \quad \text{if } d < d_{\max}, Y'(t) < Y'_S \text{ and } Y(t) < Y_R, \\ \text{else } d = d_{\max}. \tag{6}$$

$$d' = \frac{\langle Y(t) - Y'_0 \rangle_+}{Y'_c}, \quad \text{if } d' < d_{\max}, Y'(t) < Y'_S \text{ and } Y(t) < Y_R, \\ \text{else } d' = d_{\max}. \tag{7}$$

The meanings of the above used quantities are given in Tab. 1.

Table 1. Description of damage parameters

Y_C	$\left[(\text{Pa})^{1/2}\right]$	critical shear damage limit value
Y_0	$\left[(\text{Pa})^{1/2}\right]$	initial shear damage threshold value
Y'_C	$\left[(\text{Pa})^{1/2}\right]$	critical transverse damage limit value
Y'_0	$\left[(\text{Pa})^{1/2}\right]$	initial transverse damage threshold value
Y'_S	$\left[(\text{Pa})^{1/2}\right]$	brittle transverse damage limit of the fiber-matrix interface
Y_R	$\left[(\text{Pa})^{1/2}\right]$	elementary shear damage fracture limit
$d_{\max}$	$[-]$	maximum allowed value of d and d' ($d_{\max} < 1$)

Fiber tensile (compression) damage

The implemented law [3] implies that the fiber damage is null while $\varepsilon_{11} < \varepsilon_i^{ft}$, where ε_i^{ft} corresponds to the initial longitudinal fiber tensile damage threshold strain. Then the tensile fiber damage d^{ft} grows linearly between $\varepsilon_i^{ft} < \varepsilon_{11} < \varepsilon_u^{ft}$ where ε_u^{ft} is the ultimate longitudinal fiber tensile damage strain. When this ultimate value is reached, the fiber tensile damage d^{ft} reaches the ultimate damage, d_u^{ft}. After this point the tensile damage grows asymptotically towards $d^{ft} = 1$ (see Fig. 4).

5. PAM-Crash impact model validation

The PAM-Crash software includes many types of material models for composite materials as well as many element types. The Ladevèze model is chosen in this work for the analysis of impact behaviour together with shell elements. The properties of such model need to be firstly validated by experimental measurement for the chosen material. Therefore, a comparison of experimental measurement and numerical simulation in PAM-Crash is carried out. Investigated is transverse impact of metal impactor on composite plate (made of EHKF420-UD24K-40).

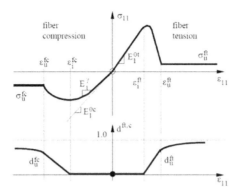

Fig. 4. Fibers tensile and compressive damage

Experimental measurements

A steel impactor was dropped from the height of $h = 300$ mm on a composite plate which was clamped along one edge was measured. A transverse displacement of the plate $u(t)$ was measured directly under the point of impact, 10 mm from the edge of the plate, see Fig. 5. The displacement was measured by a laser sensor optoNCDT by Micro-Epsilon. The experimental setup scheme is illustrated in Fig. 6.

The failure analysis of composite materials was the next comparison of the experimental measurement with numerical simulation [2]. Three stripes were made. The stripes properties were: length 105 mm, thickness 1.05 mm, width 15, 20, 25 mm and fiber orientation 0 (A00), 45 (A45) a 90 (A90). The comparison of experimental and numerical values of failure force are shown in Table. 4 bellow.

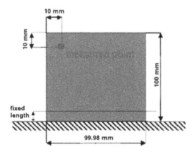

Fig. 5. Basic dimensions of composite plate

Fig. 6. Experimental measurement: a) experimental schema, b) actual photograph from measurement (1 – tube fixture, 2 – impactor guide tube, 3 – composite plate, 4 – optoNCDT laser sensor)

Table 2. Plate properties

$E_L = 109.4$ GPa
$E_T = 7.7$ GPa
$G_{LT} = 4.5$ GPa
$\nu = 0.28$
fixed length = 6 mm
plate thickness = 0.8 mm
$\varrho = 1\,579.07$ kg/m^3

Table 3. Impactor properties

$E = 210$ GPa
$\varrho = 7\,800$ kg/m^3
$\nu = 0.3$
diameter $d = 4$ mm
length $l = 40$ mm

Table 4. Comparison of failure forces

Force [kN]	A00	A45	A90
Experiment	33.28	1.35	1.11
Numerical simulation	32.722 7	1.321 8	1.196 8

The composite plate made of carbon/epoxy (C/E) material consists of four layers of unidirectional composite with orientations $[0, 90, 90, 0]$. The total measured thickness is 0.84 mm. The dimensions of the plate including the measured point are given in Fig. 5. The characteristics of the plate are shown in Table 2, where the index L denotes the longitudinal direction and the index T means the transverse direction. The steel impactor is displayed in Fig. 7 and its basic material properties and dimensions are mentioned in Table 3. To eliminate errors seven measurements were carried out. The acquired time-dependent displacement curves are displayed in Fig. 8.

Fig. 7. Impactor   Fig. 8. Comparison between experimental curves and numerical simulation

Numerical simulations

The numerical model was firstly calibrated by means of modal analysis and calculation of free oscillations of the composite plate from an experimental measurement [1]. The following parameters are in Table 2. Material parameters of the steel impactor were chosen according to Table 3. The movement of the impactor was restricted by boundary conditions only to the direction normal to the plate. The resulting displacement curve obtained by numerical simulation is compared with the experimental curves in Fig. 8.

From the above-mentioned results it can be seen that there is a good agreement between experimental measurements and numerical simulation.

6. Design of new composite bumper reinforcement

In this part new composite bumper reinforcement is proposed. In the next step the stiffness analysis is performed. It is compared with an original steel bumper reinforcement by means of stiffness analysis. The aim was to keep the basic shape of the bumper reinforcement the same as in the case of the steel bumper – it means the length, height and curvature of bumper are similar. The goal is to achieve the same, eventually better, mechanical behaviour of the bumper reinforcement because the steel reinforcement shows fracture during RCAR test as discussed below. Afterwards, a large strain damage analysis and a weight comparison of selected bumper reinforcements are performed and the corresponding weights are compared. The connection between the reinforcement and the deformable element is not investigated. The connection is rigid and realized by so-called tied elements.

6.1. Stiffness analysis

For this stiffness analysis, four different simple geometry bumper reinforcement profiles were proposed as shown in Fig. 9.

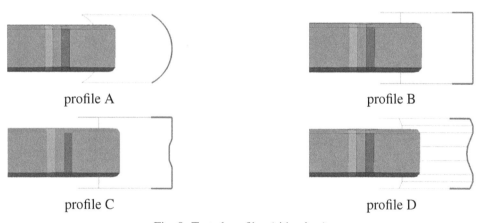

profile A   profile B

profile C   profile D

Fig. 9. Tested profiles (side view)

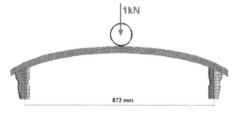

Fig. 10. Stiffness analysis

The stiffness test is displayed in Fig. 10. Steel cylindrical "impactor" is pushed into the bumper reinforcement with a force of 1 kN. The displacement of the bumper reinforcement is evaluated under the steel cylindrical position. The material used for bumper reinforcement is again EHKF420-UD24K-40 (C/E). Values of the Ladevèze model parameters were set as shown in Tab. 5.

Table 5. Values of Ladevèze model parameters.

Y_C	$[(\text{GPa})^{1/2}]$	0.219	Y_R	$[(\text{GPa})^{1/2}]$	0.204 9
Y_0	$[(\text{GPa})^{1/2}]$	0.219	ε_i^{ft}	$[-]$	0.019 5
Y_c'	$[(\text{GPa})^{1/2}]$	0.204 9	ε_u^{ft}	$[-]$	0.019 5
Y_0'	$[(\text{GPa})^{1/2}]$	0.204 9	d_{max}	$[-]$	0.98
Y_S'	$[(\text{GPa})^{1/2}]$	0.011 700 42			

At first, the lay-up of the composite structure was proposed as $[0^1, \pm30, 90, \mp30, 0]$ and the total composite thickness was 6 mm. This composite structure did not reach stiffness improvement in comparison with original steel bumper reinforcement (see [1]). The profiles A and C showed low stiffness values and, therefore, they were not taken into consideration in the next simulations. To further improve the designs, the composite structure lay-ups of both profiles B and D were modified to $[0, \pm30, \mp30, 90, \pm30, \mp30, 0]$ in the next step and so the new total composite thickness was 7.6 mm. The new results are displayed in Fig. 11. It can be seen that the modified profiles are now stiffer than the original steel structure. The complete results from stiffness analysis can be found in [1].

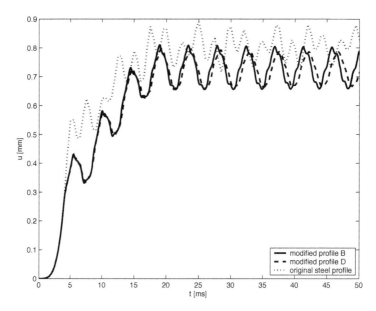

Fig. 11. Position of impactor in time

6.2. Damage analysis

The testing was realized with a rigid barrier, which was pushed against the bumper reinforcement and the deformable element, (see Fig. 12) by a prescribed displacement.

The parameters of the simulation barrier are equal to the barrier used in RCAR tests as introduced in section 2.2, i.e., rigid barrier with an inclination of 10 degrees, shift of the barrier is 40 % of the maximum width of the car without mirrors. The results are displayed in Fig. 13.

---

[1] bumper reinforcement longitudinal direction

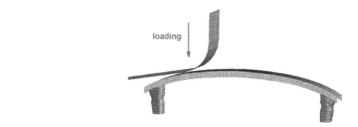

Fig. 12. Damage test

profile B   profile D

Fig. 13. Damage of composite reinforcement in simulation time 20 ms

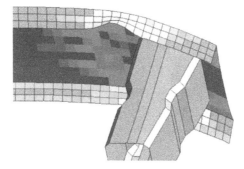

Fig. 14. The fracture initiation on the edge of modified profile B

The analysis revealed fracture on the edge of the profile B, see Fig. 14, which does not demonstrate a good damage behavior of the bumper reinforcement. The modified profile D proved to be the best of all tested profiles. The maximum value of damage in the simulation is about 60 % (maximum d_{max} value in the whole model). This means that the loading is still transferred to the deformable element (see Fig. 2) which is necessary feature for a well-designed bumper reinforcement. The original steel bumper reinforcement also shows certain fracture on the edge as shown in Fig. 15. Therefore, the composite profile D proved to be a suitable replacement for the original steel structure from both the stiffness and damage point of view.

before fracture   fracture

Fig. 15. Fracture on the original steel bumper reinforcement

The main advantage of the new bumper is the mass reduction. Mass comparison between original steel bumper reinforcement and composite bumper reinforcements is displayed in Tab. 6. The mass was calculated from elements volume and material density. The comparison shows a great mass reduction of the bumper reinforcements with use of the composite materials – up to 78 % in the case of best profile D.

Table 6. Bumper reinforcement mass comparison

Material	Mass [kg]
original steel	8.89
composite profile A	1.14
modified composite profile B	1.63
composite profile C	1.45
modified composite profile D	1.94

7. Conclusions

The work is dedicated to the simulation of car bumper behavior according to prescribed safety procedures. The main part of the work is focused on the design of a new composite bumper reinforcement with the aim to maintain or improve its mechanical properties while reducing the mass. The numerical models for PAM-Crash software using shell elements and Ledevèze model for the composite material are validated by experimental measurements. Afterward, four composite profiles are proposed and tested using stiffness analysis performed according to RCAR specification. After modification of the material structure, two profiles are found to have greater stiffness that the original steel structure and they are selected for latter damage analysis. The damage analysis proved one of the profile to be a suitable replacement for the original part as it showed no fracture and, moreover, a great mass reduction could be achieved with the use of the composite structure.

However, the matter of price and damage of the reinforcement due to composite material delamination must be further inspected. This problem has not been involved in the numerical model yet. The connection between the reinforcement and the deformable element was also not resolved in this work since it is rather a technological issue.

Acknowledgements

The work has been supported by the research project MSM 4977751303 and research project GA AV IAA200760611.

References

[1] Kleisner, V., Design of composite car bumper, Diploma thesis, University of West Bohemia, 2008. (in Czech)
[2] Laš, V., Zemčík, R., Progressive damage of unidirectional composite panels, Journal of Composite Materials, Vol. 42, No. 1, pp. 25–44, 2008.
[3] PAM-CRASH 2007. *Solver Notes Manual*. ESI-Group, Paris.
[4] Research Council for Automobile Repairs. *Rcar homepage* [on-line]. [cit. 2008-20-05]. URL: <http://www.rcar.org/index.htm>.
[5] The official site of the European New Car Assessment Programme. *Front impact* [on-line]. [cit. 2008-20-05]. URL: <http://www.euroncap.com/tests/frontimpact.aspx>.
[6] Vlk, F., Car body : ergonomics, biomechanics, passive safety, collision, structure, materials. Brno, Publisher Vlk, 2000, 243 pp., ISBN 80-238-5277-9. (in Czech)

6

A refined description of the crack tip stress field in wedge-splitting specimens – a two-parameter fracture mechanics approach

S. Seitl[a,*], Z. Knésl[a], V. Veselý[b], L. Řoutil[b]

[a]Institute of Physics of Materials, Academy of Sciences of the Czech Republic, v.v.i. Žižkova 22, 616 62 Brno, Czech Republic
[b]Institute of Structural Mechanics, Faculty of Civil Engineering, Brno University of Technology, Veveří 331/95, 602 00 Brno, Czech Republic

Abstract

The paper is focused on a detailed numerical analysis of the stress field in specimens used for the wedge splitting test (WST) which is an alternative to the classical fracture tests (bending, tensile) within the fracture mechanics of quasi-brittle building materials, particularly cementitious composites. The near-crack-tip stress field in the WST specimen is described by means of constraint-based two-parameter fracture mechanics in the paper. Different levels of constraint in the vicinity of the crack tip during fracture process through the specimen ligament are characterized by means of the T-stress. Two basic shapes of WST specimen – the cube-shaped and the cylinder-shaped one – are investigated and the determined near-crack-tip stress field parameters are compared to those of compact tension (CT) specimens according to the ASTM standard for classical and round geometry. Particular attention is paid to the effect of the compressive component of the loading force (complementing the splitting force) acting on the loaded side of the specimen and its reaction from the opposite part of the specimen on the stress field in the cracked body. Several variants of boundary conditions on the bottom side of the specimen used for this kind of testing procedure are also considered. The problem is solved numerically by means of the finite element method and results are compared with data taken from the literature.

Keywords: wedge splitting test, crack tip stress field, two-parameter fracture mechanics, fracture parameters, numerical simulation

1. Introduction

For determination of the fracture-mechanical parameters of quasi-brittle materials used commonly in the building industry, particularly cement-based composites such as concretes and mortars, specific experimental tests on notched specimens under bending (e.g. three- or four-point bending of notched beams or cylinders), tension (e.g. single or double edge notched or dog-bone specimens) or eccentric compression (e.g. double edge notched cubes or prisms and round notched cylinders) are usually performed [5, 15, 34, 39].

However, the influence of the bended specimen's own weight at the performed test's record is significant and cannot be ignored, especially in the case of materials with low tensile strength, e.g. early-aged concretes and mortars [28]. Moreover, a substantial portion of the testing specimen's volume remains elastic and does not directly participate in the failure test, which causes an unnecessary increase in the material required for preparation of the testing specimen. There is either a superfluous amount of the fresh concrete/mortar necessary for the casting of the

*Corresponding author. e-mail: seitl@ipm.cz..

testing specimens in the case of those specimens prepared simultaneously with the structure's realisation or an immoderate demand for the volume of material necessary to be taken from already existing structures.

The performance of the tensile tests is, on the other hand, very demanding on the experimental equipment, which encounters e.g. the need for high stiffness of the testing machine and fittings (e.g. [5, 40, 44]) as well as the necessity of a sophisticated apparatus controlling the loading rate through feed-back signal from gauges measuring displacements or crack opening displacements (e.g. [1]).

Eccentric compression is disadvantageous, especially for the reason that also applies to some cases of both above-mentioned loading configurations; scilicet, a substantial amount of elastic energy stored in the specimen and the testing machine (frame, fittings) is released suddenly at the start of the crack/notch propagation, which causes an instable fracture [41].

A convenient alternative to the bending or tensile tests is presented by a wedge splitting test (WST) proposed by [24] and later developed in [7] and other works. The WST is an adaptation of the common compact tension (CT) test which eliminates the disadvantages stemming from the usually insufficient toughness of the fittings between the CT specimen and the testing machine (cumulating of elastic energy resulting in lower test stability). A compressive load induced by the loading device is transformed into a tensile loading opening the initial notch via special testing arrangement based on the wedge mechanism (see fig. 1a. An ordinary electromechanical testing machine with a constant actuator displacement rate can be used and no sophisticated test stability control apparatus (i.e. closed loop control unit with e.g. crack tip opening displacement as a feedback signal) is necessary. A cardinal advantage of this arrangement is its very high stiffness in comparison to the loading chain of the common CT test or other, especially tensile, tests.

As is obvious from fig. 1, the test specimens can be prepared either from standard cube, beam or cylinder-shaped specimens cast into standard moulds, or as prismatic or cylindrical specimens taken from existing structures by sawing or core-drilling, respectively. The WST is extensively used for various experimental studies and recently an increase in usage of the testing method has been registered (e.g. [9, 14, 19, 25, 28, 38, 46, 47, 48]).

The determination of fracture-mechanical properties of materials from records of fracture tests is conditioned by a proper fracture-mechanical description of the test in question. In the case of common testing geometries relevant information can be found summarized in the classical works in the field of fracture mechanics (e.g. [2, 5, 15, 29, 34]), handbooks ([26, 36]) or other works [20]. Fracture parameters for the WST are not so widely reported. Stress intensity factors (SIF, K-factor) for particular variants of the WST (cube-shaped specimens) can be found in literature, e.g. [10, 30]. However, they are utilizable within a classical (single-parameter) fracture mechanics approach only, which is inaccurate (insufficient) for application in the case of quasi-brittle fracture. Approximate expressions of the terms of Williams' series [45] approximating the stress field in the cracked body up to the order of 5 for particular dimensions of cube-shaped WST specimens were introduced for the first three coefficients in [18] and further developed for five coefficients in [16]. These results cover a considerably wide range of exploitable dimensions of cube-shaped WST specimens with two types of boundary conditions in the area of application of the load to the specimen by the wedge mechanism. However, they are valid only for a limited range of variants of support configuration on the bottom side of the specimen.

In this paper, the numerical analysis of stress field for wedge splitting geometry in the framework of two-parameter fracture mechanics is introduced. The stress intensity factor K and the T-stress for cube- and cylinder-shaped WST specimens are determined and they are compared

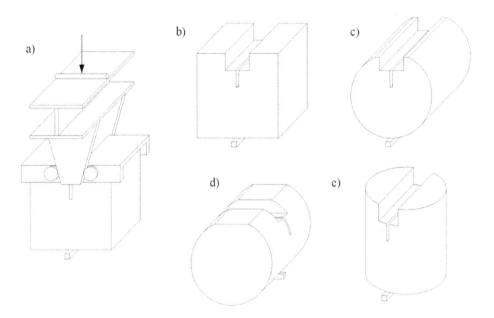

Fig. 1. Wedge splitting test geometry: sketch of stiff fixtures transferring the load from the testing machine to the specimen (a), several variants of specimen shapes applicable for WST prepared from standard cube (b) and cast or core-drilled cylinder (c, d, e) (inspired by [24])

to results corresponding to the similar testing configuration – the classical and round CT specimens. The effects of different boundary conditions on both the loaded and the supported side of the WST specimen on the calculated fracture parameters are examined. The influence of friction of the roller bearing on the values of fracture parameters (K, T-stress) is briefly discussed. The present paper follows on from and elaborates the previous studies of the authors on this subject [31, 32, 33].

This paper is divided into several sections. After this introduction the motivation of the work is clarified with particular definition of the objectives of the analyses performed. Next, the theoretical background as a basis for the calculations conducted is briefly explained. A section devoted to modelling details is then followed by analysis and discussion of the results. The importance of the research introduced with substantial conclusions is stated in the last section.

2. Motivation, analysis objective definition

Knowledge of the stress field in the cracked specimen, or at least its accurate enough estimation, provides the possibility of constructing the size and shape of the fracture process zone (FPZ) characteristic for the fracture of quasi-brittle cementitious composites, as is proposed by [42, 43]. This method is based on multi-parameter linear elastic fracture mechanics and classical non-linear models for quasi-brittle fracture. A technique of this kind enables the specification of the energy dissipated in the FPZ evolving at the macroscopic crack tip during fracture process in quasi-brittle materials to its volume, which might contribute to a more precise determination of the fracture-mechanical parameters of quasi-brittle materials. The numerical analysis proposed in the paper has been motivated particularly by this issue.

The numerical study consists of three parts which are described in following sections in detail.

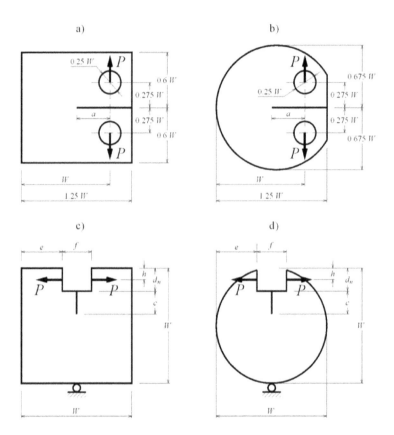

Fig. 2. The geometry of the compact tension test with classical (CT – a) and round (RCT – b) compact tension specimen; the geometry of wedge-splitting test with cube-shaped (c) and cylinder-shaped (d) specimens

2.1. Study I

In this study the characteristics of the near-crack-tip stress field in the cube- and cylinder-shaped WST specimen (figs. 2c and 2d, respectively) are compared to those of the classical and round CT specimen (figs. 2a and 2b, respectively) that are usual (standard) testing specimens in the case of fracture testing of metals [4]. The CT geometry was chosen due to its similarity to the WST geometry from both the specimen's shape and the imposed load point of view. In this study the loading of the WST specimens was considered in a simplified way – the splitting component of the loading force[1] was only assumed.

For numerical calculations of both geometries the value $W = 100$ mm was used. Other dimensions of the WST geometries were equal to $e = 35$ mm, $f = 30$ mm, $h = 10$ mm, $d_n = 20$ mm (the dimensions are indicated in fig. 2). The initiation notch of length a and c for CT and WST geometry, respectively, was modelled as a crack where their values varied so that parameter α indicating the relative crack length lies in the interval $\langle 0.2, 0.8 \rangle$. In the present numerical study the relative crack length α is defined as a ratio of the effective crack length, i.e. the distance from the point of the force application to the crack tip, and the effective specimen

---

[1]The loading force introduced by the testing machine is decomposed through the wedge into the splitting and the compressive component. As the angle of the wedge ranges usually from $10°$ to $15°$, the value of the compressive component in relation to the splitting on is very low.

width, i.e. the distance from the point of the force application to the end of the specimen:

$$\alpha = \frac{a}{W} \qquad (1)$$

for CT and RCT specimens, see figs. 2a and 2b, and

$$\alpha = \frac{c + (d_n - h)}{W - h} \qquad (2)$$

for cube- and cylinder-shaped WST specimens, see figs. 2c and 2d.

2.2. Study II

Investigation of the stress field in the WST specimens has followed the specification of the boundary conditions in the area of introducing of the load – the component of the loading force inducing the compression of the specimen was taken into account. Both expected components of the loading force in the case of WST specimen are shown in fig. 3. The applied splitting force P_{sp} (acting horizontally in the figure) is related to the compressive load P_v (vertical), see e.g. [30, 35], as:

$$P_{\mathrm{v}} = \frac{1}{2} P_{\mathrm{sp}} k, \qquad (3)$$

where

$$k = \frac{2 \tan \alpha_{\mathrm{w}} + \mu_{\mathrm{c}}}{1 - \mu_{\mathrm{c}} \tan \alpha_{\mathrm{w}}}. \qquad (4)$$

Symbol α_{w} represents an angle of the wedge and μ_{c} refers to friction in the roller bearings. A single support on the axis of symmetry presents the only boundary condition on the specimen back-face (see fig. 3).

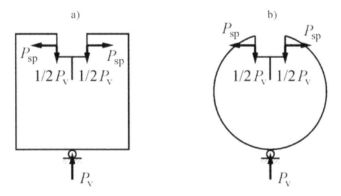

Fig. 3. The cube- and cylinder-shaped WST specimens – an indication of the loading determined by the boundary conditions

The influence of the friction in roller bearings which transfer the load from the wedge to the specimen was also investigated. The coefficient of friction μ_{c} usually varies between 0.001 and 0.005 for bearings used within the experimental setup, see e.g. [13]. In our case two values of coefficient μ_{c} are used as the lower and upper bound to cover all possible values of the friction coefficient. The first one was set to 0; it means that there is no friction in the bearings, and second one to 0.005, which represents the maximal friction in the roller bearings. The dimensions of the WST specimens were considered the same as in the previous study.

2.3. Study III

The next study was focused particularly on the cube-shaped WST specimens. Precise representation of the loading (both components of the loading force) was taken into consideration and the influence of the boundary conditions on the bottom side of the specimen was chiefly investigated.

The study supplements the work by Karihaloo et al. [16] in this aspect. In figs. 4a and 4b schemes of testing configuration analysed in [16] are depicted. Schemes in figs. 4c up to 4e represent variants of boundary conditions on which own computations were conducted. On the other hand, only the values of the stress intensity factor K and the T-stress along the crack propagation through the specimen ligament were computed in contrast to first 5 terms of the Williams series presented in [16], as is noted above. Another improvement covered by the proposed paper in comparison to the results published in [16] is the investigation of the influence of the vertical component of the loading force on the values of the near-crack-tip stress state parameters (compare figs. 4a with 4c or 4b with 4e, respectively). This issue is partly also the subject of the study described in the previous section. The specimen dimensions correspond to the previous studies and are indicated in fig. 4.

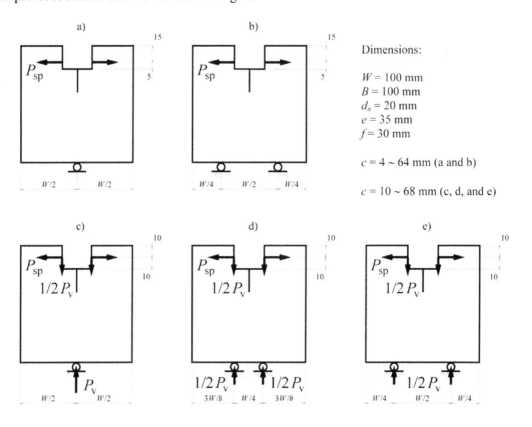

Fig. 4. Considered specimen dimensions, components of loading force and variants of supports: (a and b) Karihaloo et al. [16], (c, d, and e) own computations

3. Theoretical background

According to two-parameter fracture mechanics approach which uses T-stress as a constraint parameter, the stress field around the crack-tip of a two-dimensional crack embedded in an isotropic linear elastic material subjected to normal mode I loading conditions is given by the following expressions [45, 2]:

$$
\begin{aligned}
\sigma_{xx} &= \frac{K_I}{\sqrt{2\pi r}} \cos\left(\frac{\theta}{2}\right)\left[1 - \sin\left(\frac{\theta}{2}\right)\sin\left(\frac{3\theta}{2}\right)\right] + T \\
\sigma_{yy} &= \frac{K_I}{\sqrt{2\pi r}} \cos\left(\frac{\theta}{2}\right)\left[1 + \sin\left(\frac{\theta}{2}\right)\sin\left(\frac{3\theta}{2}\right)\right] \\
\tau_{xy} &= \frac{K_I}{\sqrt{2\pi r}} \cos\left(\frac{\theta}{2}\right)\sin\left(\frac{\theta}{2}\right)\cos\left(\frac{3\theta}{2}\right)
\end{aligned}
\tag{5}
$$

where r and θ are the polar coordinates and x and y are the Cartesian coordinates, both with origins at the crack tip. K_I is the stress intensity factors (SIF) for mode I, σ and τ are the normal and shear stress, respectively. The T term in the first line of equation (6) is the elastic T-stress which is the second constant term corresponding to a uniform parallel stress $\sigma_{xx} = T$. Thus, in two-parameter based fracture mechanics, the stress field is expressed by means of the two parameters, the stress intensity factor K_I and the T-stress (see e.g. [20, 27]).

The SIF and the T-stress according to [23] may be normalized expressed by stress biaxiality ratio B as follows:

$$
B_1 = \frac{K_I}{K_0}, \qquad \text{where} \qquad K_0 = \frac{P_{sp}}{t\sqrt{W}}
\tag{6}
$$

and

$$
B_2 = \frac{T\sqrt{\pi a}}{K_I},
\tag{7}
$$

where P_{sp} is the horizontal splitting loading force, t is the thickness, W is the fundamental dimension (specimen width) and a represents the crack length. In the present numerical study the crack length a is defined as a distance from the point of the force application to the crack tip, see section 2.1.

4. Modelling

The material input data for the concrete used in the numerical simulation are chosen as follows: Young modulus $E = 44\,000$ MPa and Poisson's ratio $\nu = 0.2$.

The stress intensity factor K and the T-stress values were computed by means of the finite element method either using direct techniques [49] or quarter-point crack-tip elements for their determination ([37, 11]). Generally, the direct methods need extreme mesh refinement close to the crack tip in comparison with the method employing quarter-point elements. For direct method the estimation of the fracture parameters is derived directly from the singular stress description, see eq. (5).

As a first step, the 2D finite element method solution is employed to the CT and RCT specimens to verify the accuracy of used numerical model. Subsequently the above-mentioned methods are applied to fracture parameters calculations for both shapes of WST specimens. The commercial finite element program ANSYS is used to analyze all the models presented here.

Note that the geometries are symmetric including both the specimen shapes and the loading conditions; therefore only one half of each specimen was modelled. Different finite element

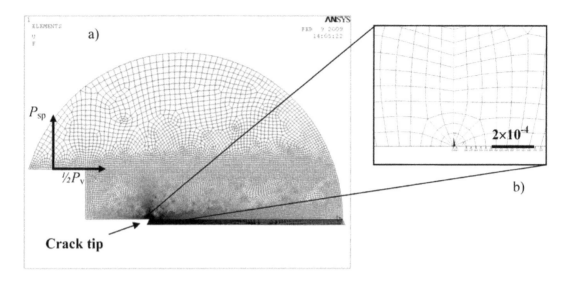

Fig. 5. The finite element mesh used in the simulations: a) One half of the cylinder shaped WST specimen used for FEM calculation, b) detailed view of the small region near the crack tip (a quarter-point crack-tip element was used)

models are constructed for each of the a/W ratios investigated. A typical finite element mesh used in the computations is shown in fig. 5 together with a detailed view of the small region near the crack tip. The size of the smallest element in the crack tip is 5×10^{-5} mm.

5. Results and discussion

The results obtained from the numerical study are divided into following three subsections correspondingly to section 2.

5.1. Study I

Numerical results comparing the normalized values of the SIF (i.e. B_1) and T-stress (i.e. B_2) for the cube- and cylinder-shaped WST specimens (see figs. 2c and 2d, respectively) with those of the classical and round CT specimens (see figs. 2a and 2b, respectively) are plotted in fig. 6. The values of the B_1 and B_2 for the classical and round CT specimens can be found in literature, see e.g. [20].

The compact tension geometry from the ASTM standard [4] is similar, in some geometrical aspects; to the wedge splitting geometry and it is possible to expect similar values of the normalized stress intensity factor B_1 related to SIF by eq. (6). The dependences of normalized stress intensity factor B_1 as a function of α for four considered test geometries are presented in fig. 6 left. The graph shows that the normalized SIF values are similar for corresponding specimen shapes of wedge splitting and compact tension geometry (cubic-shaped WST and CT, cylindrical WST and RCT). This fact is obvious especially for long cracks ($\alpha > 0.5$), where the influence of the loading setup to the stress state within the specimen can be supposed to vanish.

The normalized T-stress values B_2 defined as a function T-stress in eq. (6), are plotted in fig. 6 right. According to the results obtained, it can be seen that the $B_2(\alpha)$ functions for the WST geometry vary from negative to positive values for both specimen shapes. This is contrary to the compact tension specimen (CT and RCT), where the B_2 values are positive for each α.

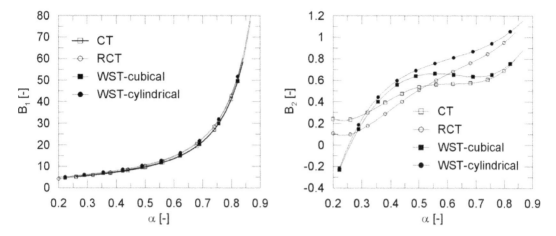

Fig. 6. Normalized values of the stress intensity factor K, B_1 (left), and T-stress, B_2 (right), for four considered specimen shapes

5.2. *Study II*

Numerical results of this study show the influence of the compressive load P_v (vertical) defined in eq. (4), see fig. 3. As the second step in this section the influence of the friction in roller bearings which transfer the load from the wedge to the specimen was investigated.

The numerically calculated values of the normalized SIF (i.e. B_1) and T-stress (i.e. B_2) for cube- and cylinder-shaped WST specimens are given in fig. 7 (left) and 7 (right) and fig. 8 (left) and 8 (right), respectively. In these figures there are always three curves plotted: *i)* marked as WST – the curve for cube-/cylinder-shaped WST loaded only by splitting force, *ii)* WST_P – cube-/cylinder-shaped WST loaded by both components of the loading force, i.e. the splitting force P_{sp} and the vertical compressive force P_v, without considering friction in roller bearings (i.e. $\mu_c = 0$), and *iii)* WST_P_μ – the same as in *ii)* with considering the maximal potential value of friction coefficient ($\mu_c = 0.005$).

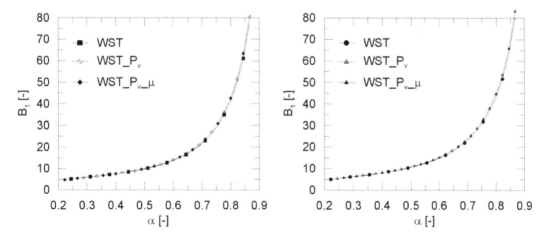

Fig. 7. Normalized values of the stress intensity factor K, B_1, for cube-shaped (left) and cylinder-shaped (right) WST specimen

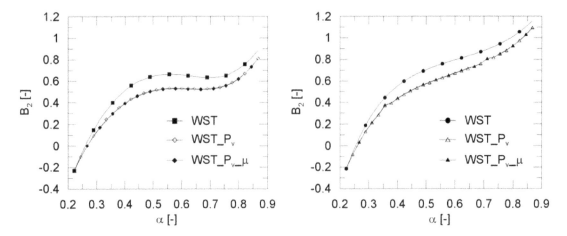

Fig. 8. Normalized values of the T-stress, B_2, for cube-shaped (left) and cylinder-shaped (right) WST specimen

The results for cube-shaped WST specimen are shown in fig. 7 and 8 on the left. The B_1 as a function of α have the same trend for all (WST, WST_P and WST_P_μ) curves, see fig. 7 left. A slight difference of the B_1 values (weak increase) can be seen for analyses considering the vertical compressive force in comparison to the analysis neglecting the effect of the bottom support, however, only for $\alpha > 0.7$. The influence of the bottom support (differences between the WST and WST_P curves) is more pronounced than the influence of the friction in the roller bearings (differences between WST_P and WST_P_μ curves).

The normalized values of second parameter – T-stress, expressed by B_2, as a function of α are presented in fig. 8. The influence of the vertical compressive load (WST_P) plays a dominant role on the parameter B_2 (see differences between WST and WST_P curve), the influence from the friction in roller bearings is insignificant. The WST_P and WST_P_μ curves are similar and the error introduced by neglecting the friction in bearings (even at its maximal value) is less than 2 %, which is in agreement with recommendations from the literature [30, 35].

The results for the cylinder-shaped WST specimen are shown in fig. 7 and 8 on the right. The graphs shows that the normalized stress intensity factor values B_1 are similar for all three studied cases of wedge splitting test, similarly to the cube-shaped WST. This fact is especially obvious for short cracks ($\alpha < 0.6$), where the influence of the loading set-up to the stress state within the specimen can be regarded as negligible.

The normalized T-stress, i.e. $B_2(\alpha)$ functions, vary from negative to positive values for all loading cases. The influence of the vertical compressive load plays a dominant role in the parameter B_2, as in the case of the cube-shaped WST, the change due to the friction in roller bearings is also insignificant.

5.3. Study III

The final study focused on the cube-shaped WST specimen and covered the influence of alternative supports, see figs. 4c (WST_P$_v$), 4d (WST_P$_v$-1/4), 4e (WST_P$_v$-1/8). The numerically calculated values of the normalized stress intensity factor K (B_1) and T-stress (B_2) are given in figs. 9 left and 9 right, respectively.

The normalized SIF values, B_1, are similar for WST cube-shaped specimen without vertical load and three specimens including vertical load with various supports (a single central line

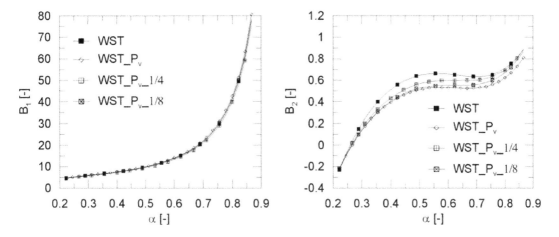

Fig. 9. Normalized values B_1 of the stress intensity factor K_I (left) and normalized values B_2 of the T-stress (right) for considered specimen dimensions and boundary conditions

support and two line support), as shown in the fig. 9 and it is possible to conclude that the influence of various supports on values of SIFs is not significant.

The normalized T-stress values from the conducted FEM analysis, expressed as B_2, are plotted in fig. 9 right. According to the results obtained, it is obvious that the $B_2(\alpha)$ functions for the WST geometry vary from negative values for short cracks to positive values for relative crack lengths larger than approximately 0.25 in all cases studied. The influence of loading arrangements studied on the values of B_2 is significant especially in the interval $\alpha \in (0.3; 0.8)$. Note that the interval is commonly used within the measurement of fracture parameters in the case of the mentioned WST geometry.

6. Comparison with data from literature

As was mentioned in the introduction, Guinea et al. [10] presented solely values of SIF for WST geometry. Karihaloo & Xiao [18] compared those values were with their own results and reported differences was less than $2\,\%$. Therefore, in this article the results given by [18] (elaborated later in [16]) are compared with our presented results. These comparisons for the normalized SIF K_I (B_1) and the T-stress (B_2) are summarized in figs. 10 and 11, respectively.

In the graphs in figs. 10 and 11, there are five curves plotted:, *i)* the curve calculated for WST loaded only by the splitting force P_{sp}, see fig. 2c (marked as WST), *ii)* the curve calculated for WST loaded by both components of the loading force, i.e. the splitting force P_{sp} and the vertical compressive force P_v, see fig. 4c, (WST_P_v) *iii)* and *iv)* the curves calculated for WST loaded by both the splitting and the vertical compressive force with two supports with various distances, see fig. 4d (WST_P_v_1/4) and fig. 4e (WST_P_v_1/8), respectively, and *v)* the curve constructed from data by Karihaloo et al. [16] corresponding to the configuration depicted in fig. 4a.

It should be noted that in the article [16] the authors did not specify whether the vertical component of the load was considered in the numerical computations or not. Fig. 11 shows good agreement of the WST curve and the Karihaloo et al [16] curve, their geometry and boundary conditions are depicted in fig. 2c (WST) and in fig. 4a (Karihaloo et al. [16]). This indicates that the vertical component of the loading force was not considered in [16].

A high level of crack-tip constraint can reduce fracture toughness and, in the case of fatigue cracks, can reduce fatigue crack growth rate [21]. The knowledge of constraint enables

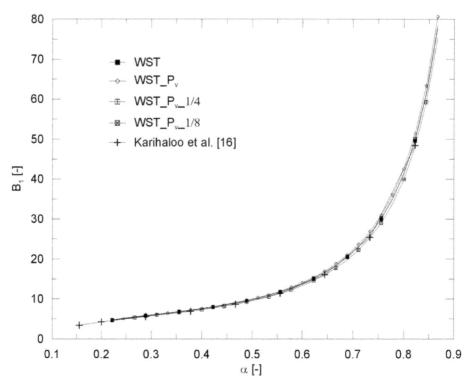

Fig. 10. Normalized values B_1 of the stress intensity factor K_I for considered specimen dimensions and boundary conditions

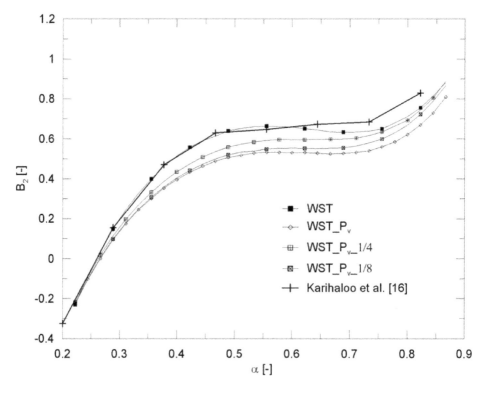

Fig. 11. Normalized values B_2 of the T-stress for considered specimen dimensions and boundary conditions

a transferability of fracture toughness values obtained on laboratory specimens to engineering structures. In general, two-parameter fracture mechanics approaches imply that the laboratory specimens must match the constraint of the structure. The two geometries must have the same B_2 values in order to transfer fracture toughness from laboratory specimens to engineering structures, see e.g. [27]. The approach of two-parameter fracture mechanics is descriptive but not predictive. From this point of view the approximation of B_2 values calculated for homogenous specimens is sufficient and can be used to characterize constraint in cases when heterogeneity not essential as well. In order to use the two parameter fracture mechanics approach in fracture analysis, the constraint for the considered wedge splitting test has to be calculated.

7. Conclusions

A finite element numerical analysis of the near-crack-tip stress fields for several variants of WST specimen shapes (cubical and cylindrical) with various boundary conditions was performed by means of a constraint-based two-parameter fracture mechanics approach. The study consists of three parts each of which was focused on selected aspects:

- i. Comparison of the cube- and cylinder-shaped WST specimens with classical CT and round CT specimens.

- ii. Consideration of the compressive component of the loading force together with a single central support on the back-face of the specimen; investigation of the influence of friction in the roller bearings.

- iii. Consideration of the compressive component of the loading force together with several variants of supports proposed in the literature.

The analysis procedures and results were compared with existing solutions from the literature. The following conclusion may be drawn from the results obtained:

- The stress intensity factors for cube-shaped WST and CT specimens and cylinder-shaped WST and RCT specimens, respectively, are similar. That holds true even for mutual comparison of cube-and cylinder-shaped WST specimens. This fact is especially pronounced for long cracks ($\alpha > 0.5$).

- The T-stress depending upon the crack depth ratio α varies from negative to positive values with increasing α for both WST specimen shapes, contrary to the CT geometry, where the T-stress is always positive.

- The influence of the geometry is much stronger on the values of T-stress than the stress intensity factors.

- The influence of the vertical compressive force, i.e. the reaction from the support at the back-face of the specimen, is much stronger on the values of T-stress than on the K-factors.

- The friction in roller bearings with usual intensity influences the results of the wedge splitting test negligibly.

- Normalized values of K-factor (B_1) are almost the same in the whole range of the relative crack length α for all studied variants of the WST boundary conditions. They are in good agreement with the results published in the literature [10, 16, 18].

- Normalized values of the T-stress (B_2) depend noticeably on both considered groups of changes of boundary conditions, i.e. both in the part of the application of the load (based on considering the compressive force component or not) and on the opposite side of the specimen (one or two supports, different distances of the supports).

- Neglecting of the compressive component of the loading force increases stress constraint at the crack tip. This tends to underestimation of the size of the zone of failure, which consequently causes an overestimation of the values of the determined fracture-mechanical parameters.

- An increase in the distance of the two supports on the bottom side of the specimen leads to an increase in the value of the T-stress.

The present results can be used for more reliable determination of fracture parameters of concrete-like materials. The utilization of the procedure for the FPZ size and shape estimation shows that knowledge of the higher order terms of the crack tip stress field is necessary [42, 43, 33], therefore attempts to compute more than the two members (i.e. K-factor and T-stress) of the Williams' series approximating the stress field in the cracked body are in preparation.

Acknowledgements

The work has been supported by the grant project of the Grant Agency of the Academy of Sciences of the Czech Republic No. KJB200410901 and by the research project of the Czech Science Foundation No. 101/08/1623.

References

[1] Akita, H., Koide, H., Tomon, M., Sohn, D., A practical method for uniaxial tension test of concrete, Materials and Structures 36 (2003) 365–371.

[2] Anderson, T. L., Fracture Mechanics: Fundamentals and Applications, Third Edition, CRC Press, 2004.

[3] ANSYS Users manual version 10.0, Swanson Analysis System, Inc., Houston, 2005.

[4] ASTM, Standard E 647–99: Standard Test Method for Measurement of Fatigue Crack-Growth Rates, 2000 Annual Book of ASTM Standards, Vol. 03. 01 (2000) 591–630.

[5] Bažant, Z. P., Planas, J., Fracture and size effect in concrete and other quasi-brittle materials. CRC Press, Boca Raton, 1998.

[6] Bednář, K., Two-parameter fracture mechanics: calculation of parameters and their meaning in description of fatigue cracks (in Czech), Ph.D. Thesis, IPM AS CR and BUT Brno, 1999.

[7] Brühwiler, E., Wittmann, F. H., The wedge splitting test, a new method of performing stable fracture mechanics test, Engineering Fracture Mechanics 35 (1990) 117–125.

[8] Červenka, V. et al., ATENA Program Documentation, Theory and User manual, Cervenka Consulting, Prague, 2005.

[9] Elser, M., Tschegg, E. K., Finger, N., Stanzl-Tschegg, S. E., Fracture behaviour of polypropylene-fibre reinforced concrete: an experimental investigation, Composite Science and Technology 56 (1996) 933–945.

[10] Guinea, G. V., Elices, M., Planas, J. Stress intensity factors for wedge-splitting geometry, International Journal of Fracture 81 (1996) 113–124.

[11] Hutař, P. Calculation of T-stress by means of shifted node method (in Czech), Proceedings of Problémy lomové mechaniky IV, Brno, 2004, 28–39.

[12] Hillerborg, A., Modéer, M., Petersson, P.-E., Analysis of crack formation and crack growth in concrete by means of fracture mechanics and finite elements. Cement and Concrete Research 6 (1976) 773–782.

[13] Interactive Engineering Catalogue: http://iec.skf.com.

[14] Ishiguro, S., Experiments and analyses of fracture properties of grouting mortars, Proceedings of Fracture Mechanics of Concrete and Concrete Structures – New Trends in Fracture, Catania, Taylor & Francis, 2007, 293–298.

[15] Karihaloo, B. L., Fracture mechanics and structural concrete, Longman Scientific & Technical, New York, 1995.

[16] Karihaloo, B. L., Abdalla, H., Xiao, Q. Z., Coefficients of the crack tip asymptotic field for wedge splitting specimens, Engineering Fracture Mechanics 70 (2003) 2 407–2 420.

[17] Karihaloo, B. L., Xiao, Q. Z., Higher order terms of the crack tip asymptotic field for a notched three-point bend beam. International Journal of Fracture 112 (2001) 111–128.

[18] Karihaloo, B. L., Xiao, Q. Z., Higher order terms of the crack tip asymptotic field for a wedge-splitting specimen, International Journal of Fracture 112 (2001) 129–137.

[19] Kim, J. K., Kim, Y. Y., Fatigue crack growth of high-strength concrete in wedge-splitting test, Cement and Concrete Research 29 (1999) 705–712.

[20] Knésl, Z. and Bednář, K., Two parameter fracture mechanics: calculation of parameters and their values, IPM of AS of Czech Republic, 1997.

[21] Knésl, Z., Bednář, K., Radon, J. C., Influence of T-stress on the rate of propagation of fatigue crack, Physical Mesomechanics (2000) 5–9.

[22] Larsson, S. G., and Carlsson, A. J., Influence of non-singular stress terms and specimen geometry on small scale yielding at crack tips in elastic-plastic material, Journal of Mechanics and Physics of Solids 21 (1973) 263–278.

[23] Leevers, P. S. and Radon, J. C., Inherent stress biaxiality in various fracture specimen geometries, International Journal of Fracture 19 (1983) 311–325.

[24] Linsbauer, H. N., Tschegg, E. K., Fracture energy determination of concrete with cube-shaped specimens, Zement und Beton 31 (1986) 38–40.

[25] Löfgren, I., Stang, H., Olesen, J. F., Fracture properties of FRC determined through inverse analysis of wedge splitting and three-point bending tests, Journal of Advanced Concrete Technology 3 (2005) 423–434.

[26] Murakami, Y., et al., Stress Intensity Factor Handbook I, II, III, Pergamon Press, Oxford, 1987.

[27] O'Dowd, N. P., Shih, C. F., Two-parameter fracture mechanics: theory and Applications, Fracture Mechanics, Philadelphia, 24 (1994) 21–47.

[28] Østergaard, L., Stang, H., Olesen, J. F., Time-dependent fracture of early age concrete, Proceedings of Non-Traditional Cement & Concrete, Brno, Brno University of Technology, 2002, 394–408.

[29] Pook, L. P., Linear Elastic Fracture Mechanics for Engineers: Theory and Applications, WIT Press, 2000.

[30] Rilem Report 5: Fracture Mechanics Test Methods for Concrete, Edited by S. P. Shah and A. Carpinteri and Hall, London, 1991.

[31] Seitl, S., Veselý, V., Řoutil, L., Numerical analysis of stress field for wedge splitting geometry, Proceedings of Applied Mechanics 2009, Smolenice, 2009, 270–278.

[32] Seitl, S., Dymáček, P., Klusák, J., Řoutil, L., Veselý, V., Two-parameter fracture analysis of wedge splitting test specimen, Proceedings of the 12th Int. Conf. on Civil, Structural and Environmental Engineering Computing, B. H. V. Topping, L. F. Costa Neves and R. C. Barros (eds), Funchal, Civil-Comp Press, 2009.

[33] Seitl, S., Hutař, P., Veselý, V., Keršner, Z., T-stress values during fracture in wedge splitting test geometries: a numerical study, submitted to conference Brittle Matrix Composites, Warsaw, 2009.

[34] Shah, S. P., Swartz, S. E., Ouyang, C., Fracture mechanics of structural concrete: applications of fracture mechanics to concrete, rock, and other quasi-brittle materials, John Wiley & Sons, Inc., New York, 1995.

[35] Skoček, J. and Stang, H., Inverse analysis of the wedge-splitting test, Engineering Fracture Mechanics 75 (2008) 3 173–3 188.

[36] Tada, H., Paris, P. C., Irwin, G. R., The Stress Analysis of Cracks Handbook, Third Edition, ASME, New York, 2000.

[37] Tan, C. L., Wang, X., The use of quarter-point crack-tip elements for T-stress determination in boundary element method analysis, Engineering Fracture Mechanics 70 (2003) 2 247–2 252.

[38] Trunk, B., Schober, G., Wittmann, F. H., Fracture mechanics parameters of autoclaved aerated concrete. Cement and Concrete Research 29 (1999) 855–859.

[39] van Mier, J. G. M., Fracture processes of concrete: Assessment of material parameters for fracture models, CRC Press, Boca Raton, 1997.

[40] van Vliet, M. R., van Mier, J. G. M., Experimental investigation of size effect in concrete and sandstone under uniaxial pension, Engineering Fracture Mechanics 65 (2000) 165–188.

[41] Veselý, V., Parameters of concrete for description of fracture behaviour (in Czech). Ph.D. thesis, Brno University of Technology, Faculty of Civil Engineering, Brno, Czech Republic, 2004.

[42] Veselý, V., Frantík, P., Development of fracture process zone in quasi-brittle bodies during failure, Proceedings of Engineering Mechanics 2009, Svratka, 288–289 + 12 p. (CD – in Czech).

[43] Veselý, V., Frantík, P., Keršner, Z., Cracked volume specified work of fracture, Proceedings of the 12th Int. Conf. on Civil, Structural and Environmental Engineering Computing, B. H. V. Topping, L. F. Costa Neves and R. C. Barros (eds), Funchal, Civil-Comp Press, 2009.

[44] Vořechovský, M., Sadílek, V., Computational modeling of size effects in concrete specimens under uniaxial tension. International Journal of Fracture 154 (2009) 27–49.

[45] Williams, M. L., On the stress distribution at the base of stationary crack, ASME Journal of Applied Mechanics 24 (1957) 109–114.

[46] Xiao, J., Schneider, H., Donnecke, C., Konig, G., Wedge splitting test on fracture behaviour of ultra high strength concrete, Construction and Building Materials 18 (2004) 359–365.

[47] Xu, S., Bu, D., Gao, H., Yin, S., Liu, Y., Direct measurement of double-K fracture parameters and fracture energy using wedge-splitting test on compact tension specimens with different size, Proceedings of Fracture Mechanics of Concrete and Concrete Structures – New Trends in Fracture, Catania, Taylor & Francis, 2007, 271–278.

[48] Wang, J. M., Zhang, X. F., Xu, S. L., The experimental determination of double-K fracture parameters of concrete under water pressure, Proceedings of Fracture Mechanics of Concrete and Concrete Structures – New Trends in Fracture, Catania, Taylor & Francis, 2007, 279–284.

[49] Yang, B., Ravi-Chandar, K., Evaluation of elastic T-stress by the stress difference method, Engineering Fracture Mechanics 64 (1999) 589–605.

7

Influence of crucial parameters of the system of an inverted pendulum driven by fibres on its dynamic behaviour

P. Polach[a,*], M. Hajžman[a], Z. Šika[b]

[a] Section of Materials and Mechanical Engineering Research, Výzkumný a zkušební ústav Plzeň s. r. o., Tylova 1581/46, 301 00 Plzeň, Czech Republic

[b] Department of Mechanics, Biomechanics and Mechatronics, Faculty of Mechanical Engineering, Czech Technical University in Prague, Technická 4, 166 07 Praha, Czech Republic

Abstract

Fibres, cables and wires can play an important role in design of many machines. One of the most interesting applications is replacement of chosen rigid elements of a manipulator or a mechanism by fibres. The main advantage of this design is the achievement of a lower moving inertia, which leads to a higher mechanism speed, and lower production costs. An inverted pendulum attached and driven by two fibres serves as a typical testing system for the investigation of the fibres properties influence on the system dynamic response. The motion of the pendulum of this nonlinear system is investigated using the **alaska** simulation tool. The influence of some parameters of the system of inverted pendulum driven by fibres has been investigated. The evaluation of influence of these crucial parameters of the system of inverted pendulum driven by fibres on its dynamic behaviour is given in this article.

Keywords: inverted pendulum, fibre, kinematic excitation

1. Introduction

Replacement of the chosen rigid elements of manipulators or mechanisms by fibres or cables [1] is advantageous due to the achievement of a lower moving inertia, which can lead to a higher machine speed, and lower production costs. Drawbacks of using such flexible elements can be associated with the fact that fibres should be only in tension [19, 20] in the course of a motion. The possible fibre modelling approaches should be tested and their suitability verified in order to create efficient mathematical models of cable-based manipulators mainly intended for the control algorithm design. Fibres are modelled using a simple force approach in almost all cases (e.g. [23]). However, in this article a more advanced model based on a point-mass approach is studied in more detail.

An inverted pendulum driven by two fibres attached to a frame (see Fig. 1) is a simplified representation of a typical cable manipulator. Real example of such a manipulator is e.g. HexaSphere redundant parallel spherical mechanism [24] (see Fig. 2), at which the replacement of rigid links by fibres is considered [20]. Functionality of fibres must be verified on this simpler example. The motion of the pendulum of this nonlinear system is investigated using the **alaska** simulation tool and an in-house software created in the MATLAB system. The influence of some parameters of the system of inverted pendulum has been investigated. The influence of

*Corresponding author. e-mail: polach@vzuplzen.cz.

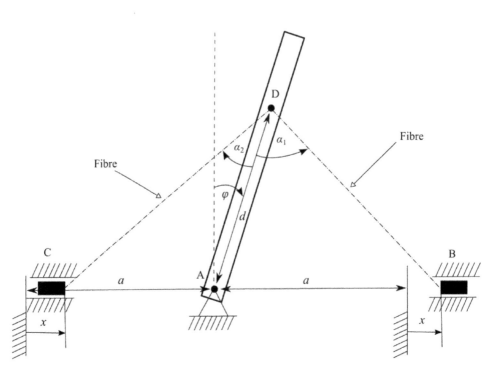

Fig. 1. Scheme of the inverted pendulum actuated by fibres

Fig. 2. HexaSphere [24]

the actuated fibres motion on the pendulum motion in the case of simultaneous harmonic excitation of fibres was investigated in [11] or [17], the influence of the phase shift in the case of non-symmetric harmonic excitation of fibres was investigated in [12]. The effect of the fibres preload on the pendulum motion was investigated in [13], the effect of the mass of the fibres on the pendulum motion was investigated in [15] and the influence of the amplitude of the harmonic kinematic excitation of fibres on the pendulum motion was investigated in [14] (all of them in the case of fibres simultaneous harmonic excitation). The evaluation of the influence of these crucial parameters of the system of inverted pendulum on its dynamic behaviour is given in this article.

As it was already mentioned, the point-mass model of the fibres is considered in the model of the system of inverted pendulum. The model of the inverted pendulum system is considered to be two-dimensional. Each fibre is discretized using 10 point masses (e.g. [11]). Each point mass is unconstrained (i.e. number of degrees of freedom is 3) in a two-dimensional model of the system. The adjacent point masses are connected using spring-damper elements. Only axial spring and damping forces are considered in these spring-damper elements. The stiffness and the damping coefficients between the masses are determined in order to keep the global properties of the fibre model based on the force approach. The validation of the point-mass model is given in [16].

2. Possibilities of fibre modelling

The fibre (cable, wire etc.) modelling [5] should be based on considering the fibre flexibility and the suitable approaches can be based on the flexible multibody dynamics (see e.g. [4, 18]). The simplest way how to incorporate fibres in equations of motion of a mechanism is the force representation of a fibre (e.g. [2]; the massless fibre model). It is assumed that the mass of fibres is small to such an extent comparing to the other moving parts that the inertia of fibres is negligible with respect to the other parts. The fibre is represented by the force dependent on the fibre deformation and its stiffness and damping properties. This way of the fibre modelling is probably the most frequently used model in the cable-driven robot dynamics and control.

A more precise approach is based on the representation of the fibre by a point-mass model (e.g. [7]). The fibre can be considered either flexible or rigid. It has the advantage of a lumped point-mass model. The point masses can be connected by forces or constraints.

In order to represent bending behaviour of fibres their discretization using the finite segment method [18] or so called rigid finite elements [22] is possible. Standard multibody codes (SIMPACK, MSC.ADAMS, **alaska** etc.) can be used for this purpose. Other more complex approaches can utilize nonlinear three-dimensional finite elements [3] or can employ the absolute nodal coordinate formulation (ANCF) elements [4, 6, 8, 18].

The approaches to the modelling of the system of inverted pendulum driven by fibres were investigated in [6, 9, 10]. Implementation of the model based on the finite rigid elements into the **alaska** simulation tool proved to be unsuitable [9]. The ANCF elements cannot be implemented in the **alaska** simulation tool, verification on this approach was carried out utilizing the MATLAB system [6, 10].

3. Inverted pendulum

As an example of the investigation of fibres behaviour an inverted pendulum, which is attached and driven by two fibres (see Fig. 1) and affected by a gravitation force, was chosen. When the pendulum is displaced from the equilibrium position, i.e. from the "upper" position, it is returned back to the equilibrium position by the tightened fibre. As it has already been mentioned, this system was selected with respect to the fact that it is a simplification of possible cable-based manipulators. In addition it was supposed that the nonlinear system of the inverted pendulum attached to a frame by two fibres could show an unstable behaviour under specific excitation conditions (e.g. [21]).

For better description of the solved problem a simple massless model is presented. The massless model is shown in Fig. 1. The used point-mass model of the fibres with lumped point masses is geometrically identical [11].

The system kinematics can be described by angle φ of the pendulum with respect to its vertical position (one degree of freedom), angular acceleration $\ddot{\varphi}$ and prescribed kinematic excitation $x(t)$. The equation of motion is of the form

$$\ddot{\varphi} = \frac{1}{I_A} \left(F_{v1} d \sin \alpha_1 - F_{v2} d \sin \alpha_2 + mg\frac{l}{2} \sin \varphi \right), \tag{1}$$

where I_A is the moment of inertia of pendulum with respect to point A (see Fig. 1), α_1 and α_2 are the angles between the pendulum and the fibres, m is the pendulum mass, F_{v1} and F_{v2} are the forces acting on the pendulum from the fibres, g is the gravity acceleration, l is the pendulum length and d is the distance from the axis in point A to the position of the attachment of fibres to the pendulum (point D). Kinematic excitation acts in the points designated B and C (see Fig. 1).

The forces acting on the pendulum from the fibre are

$$F_{v1} = \left[k_v(l_{v1} - l_{v0}) + b_v \frac{dl_{v1}}{dt} \right] \cdot H(l_{v1} - l_{v0}),$$
$$F_{v2} = \left[k_v(l_{v2} - l_{v0}) + b_v \frac{dl_{v2}}{dt} \right] \cdot H(l_{v2} - l_{v0}), \tag{2}$$

where k_v is the fibre stiffness, b_v is the fibre damping coefficient, l_{v0} is the original length of the fibres and $H(\cdot)$ is the Heaviside function. It is supposed that forces act in the fibres only when the fibres are in tension.

Actual lengths l_{v1} and l_{v2} of the fibres should be calculated in each time

$$l_{v1} = \sqrt{(d \cos \varphi)^2 + (a + x(t) - d \sin \varphi)^2},$$
$$l_{v2} = \sqrt{(d \cos \varphi)^2 + (a - x(t) + d \sin \varphi)^2}. \tag{3}$$

Kinematic excitation is given by function

$$x(t) = x_0 \sin(2\pi f t + \psi), \tag{4}$$

where x_0 is the chosen amplitude of motion, f is the excitation frequency, ψ is the phase shift (in case of symmetric excitation $\psi = 0$) and t is time.

The chosen model parameters (see Fig. 1) are: $l = 1$ m, $a = 1.2$ m, $d = 0.75$ m, $I_A = 3.288$ kg $\cdot$ m^2, $m = 9.864$ kg, $k_v = 8.264 \cdot 10^3$ N/m (stiffness), $b_v = 5 \cdot 10^{-4} \cdot k_v$ N $\cdot$ s/m (damping coefficient). Additional parameters of the point-mass models are [11–13, 17] fibre cross-section area $A_v = \pi \cdot 0.000\,001$ m^2 and fibre density $\rho_v = 4\,000$ kg $\cdot$ m^{-3} (these parameters represent the wattled steel wire — see Fig. 3). In this case mass of one fibre is 17.783 grams.

The natural frequency of the linearized system of the inverted pendulum in equilibrium position is 5.04 Hz. Thus the extreme values of pendulum angle without the parameters change appears at excitation frequency 5 Hz [11, 17].

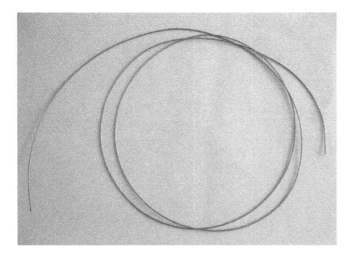

Fig. 3. Wattled steel wire

4. Simulation results

As it was already stated, the fibres preload [13], the mass of the fibres [15], the amplitude of the harmonic kinematic excitation of fibres [14] (all in the case of fibres simultaneous harmonic excitation) and the phase shift the case of non-symmetric harmonic excitation of fibres were evaluated as the crucial parameters of the system of the inverted pendulum driven by fibres [12].

Presented results are obtained using the **alaska** simulation tool. Generated nonlinear equations of motion are solved by means of numerical time integration. The simulations results presented in this article were obtained utilizing the Livermore Solver for Ordinary Differential Equation (LSODE) for stiff systems, maximum relative error that **alaska** allows at each integration step was chosen 0.000 1 and maximum absolute error that **alaska** allows at each integration step was chosen 0.000 1, too. Time step of this integration routine is variable.

Kinematic excitation amplitude was chosen $x_0 = 0.02$ m (excepting the investigation of the influence of the amplitude of the harmonic kinematic excitation of fibres on the pendulum motion [14], where it is changed in the range from $x_0 = 0.02$ m to $x_0 = 0.2$ m). Excitation frequency f was considered in the range from 0.1 Hz to 200 Hz. Some of results are given in the frequency range from 0.1 Hz to 10 Hz because the upper limit of excitation frequencies 200 Hz is too high for the practical use in manipulators.

Time histories and extreme values of pendulum angle, of the forces in the fibres and of the positions of the point masses are the monitored quantities. At excitation amplitude $x_0 = 0.02$ m maximum value of pendulum angle at quasi-static loading is $\varphi = 1.52°$; minimum value of pendulum angle at quasi-static loading is logically $\varphi = -1.52°$. Selected results of the numerical simulations are presented in Figs. 4 to 12. Simulation time is 10 seconds. It was verified that after this period the character of the system response to the kinematic excitation, apart from exceptions, does not change (e.g. [11]).

Besides the excitation frequency [11, 17] of moving fibres the pendulum motion influences all the investigated parameters (i.e. the phase shift in the case of non-symmetric harmonic excitation of fibres, the fibres preload, the mass of the fibres and the amplitude of the harmonic kinematic excitation of fibres).

Based on the obtained results it is evident that the pendulum motion is mostly influenced (besides the excitation frequency of the moving fibres) by the fibres preload [13] and by the

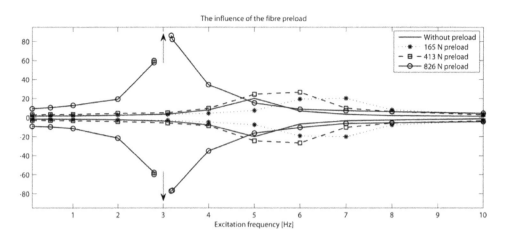

Fig. 4. Extreme values of time histories of pendulum angle φ in dependence on the excitation frequency, investigation of the influence of the fibre preload

amplitude of the harmonic kinematic excitation of fibres [14]. At the change of these parameters an unstable behaviour of the studied system was detected.

The next mentioned extreme values of the pendulum angle refer to the dependence on excitation frequency f (not to the time histories of the pendulum angle at definite excitation frequency f).

When investigating the effect of the fibres preload on the pendulum motion results at preload 165 N in fibres (i.e. at shortening by the fibres by 2 % of the free length), at preload 661 N in fibres (at shortening of the fibres by 8 % of the free length), at preload 1 033 N in fibres (at shortening of the fibres by 12.5 % of the free length) and at preload 1 099 N in fibres (at shortening of the fibres by 13.3 % of the free length) are commented in this article.

Owing to fibres preload the excitation frequency at which the pendulum angle reaches extreme values at low preload first increases (at preload 165 N in fibres up to 7 Hz) and then decreases (down to 3 Hz) at growing preload but the extreme values of pendulum angle increase — see Fig. 4. The pendulum vibration without fibres preload is more stable — in the time histories of the pendulum angle vibration damping occurs; in some cases the pendulum does not vibrate at all [11, 17]. For the first time at preload 661 N in fibres, at excitation frequency 3.83 Hz, the extreme value of pendulum angle is already greater than 90 degrees and the pendulum "oscillates" between both semiplanes defined by B and C points (see Fig. 5). From the investigated cases, e.g. at preload 1 033 N in fibres, the extreme values of pendulum angle are greater than 90 degrees at excitation frequencies in the range from 0.1 Hz to 2.6 Hz. At fibres preload 1 099 N in fibres (and higher) the extreme values of pendulum angle are greater than 90 degrees independently of excitation frequency in the whole investigated range (i.e. from 0.1 Hz to 200 Hz) [13].

The pendulum motion is influenced by the amplitude of the harmonic kinematic excitation of fibres (see Figs. 6 to 8). As it has already been stated the extreme value of pendulum angle (without the parameters change) at the investigated excitation amplitude $x_0 = 0.02$ m appears at excitation frequency 5 Hz [11–13, 17]. Owing to the amplitude of the harmonic kinematic excitation of fibres the excitation frequency at which the pendulum angle reaches the extreme value decreases (for example at excitation amplitude $x_0 = 0.06$ m down to 3.7 Hz) and on the contrary the (absolute) extreme value of the pendulum angle increases — see Fig. 6. For the first time at excitation amplitude $x_0 = 0.055$ m of fibres (at excitation frequencies in the range

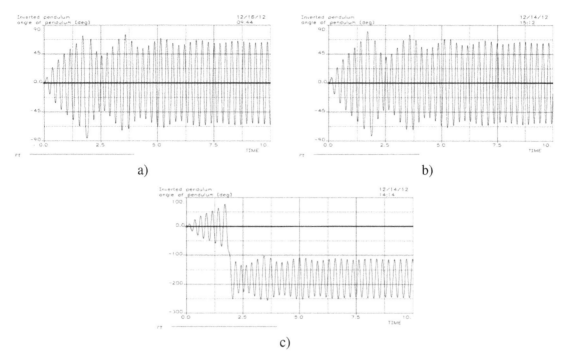

Fig. 5. Time history of pendulum angle φ, at preload 661 N in fibres, a) excitation frequency $f = 3.82$ Hz, b) excitation frequency $f = 3.84$ Hz, c) excitation frequency $f = 3.83$ Hz, investigation of the influence of the fibre preload

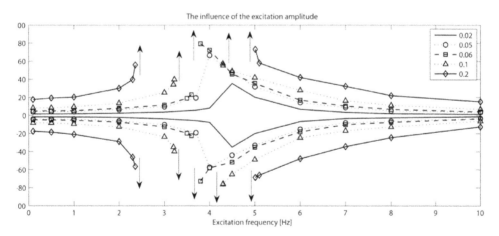

Fig. 6. Extreme values of time histories of pendulum angle φ in dependence on the excitation frequency, investigation of the influence of the excitation amplitude

from 3.76 Hz to 3.78 Hz) the extreme value of the angle is already greater than 90 degrees and the pendulum "oscillates" between both semiplanes defined by B and C points (see Fig. 8). At higher excitation amplitudes the extreme values of pendulum angle are greater than 90 degrees in some range of excitation frequencies, which extends at increasing amplitudes (e.g. at excitation amplitude $x_0 = 0.2$ m the extreme values of pendulum angle are greater than 90 degrees at excitation frequencies in the range from 2.4 Hz to 5 Hz) — see Fig. 6. Besides extreme values of pendulum angle at lower frequencies in the courses of extreme values of pendulum angle in dependence on the excitation frequency local extremes appear at higher frequencies (e.g. at

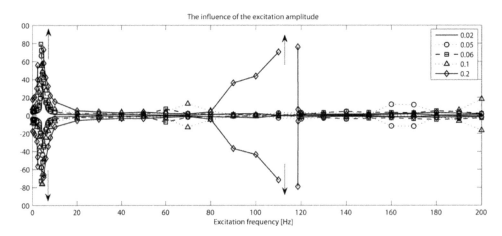

Fig. 7. Extreme values of time histories of pendulum angle φ in dependence on the excitation frequency (in the whole investigated frequency range), investigation of the influence of the excitation amplitude

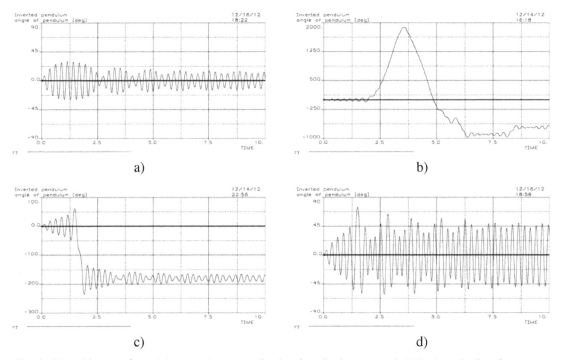

Fig. 8. Time history of pendulum angle φ, amplitude of excitation $x_0 = 0.055$, a) excitation frequency $f = 3.75$ Hz, b) excitation frequency $f = 3.76$ Hz, c) excitation frequency $f = 3.78$ Hz, d) excitation frequency $f = 3.79$ Hz, investigation of the influence of the excitation amplitude

excitation amplitude $x_0 = 0.02$ m at 40 Hz; at excitation amplitude $x_0 = 0.2$ m in the range from 111 Hz to 118 Hz the extreme value of pendulum angle is even greater than 90 degrees — see Fig. 7) [14].

Influences of the change in the mass of the fibres [15] and of the change in the phase shift in the case of non-symmetric harmonic excitation of fibres [12] do not affect the pendulum motion as significantly as the fibres preload and as the amplitude of the harmonic kinematic excitation of fibres. These parameters do not cause the unstable behaviour of the system of the inverted pendulum.

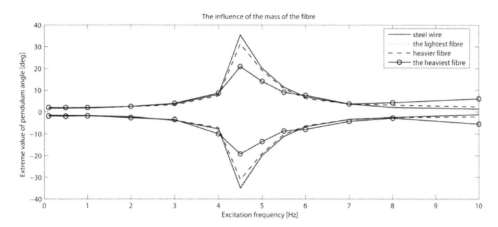

Fig. 9. Extreme values of time histories of pendulum angle φ in dependence on the excitation frequency, investigation of the influence of the mass of the fibres

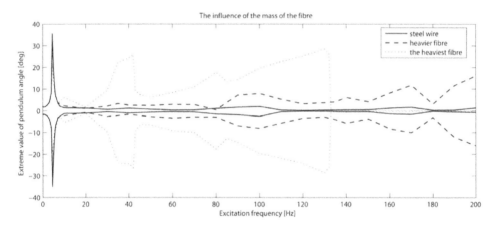

Fig. 10. Extreme values of time histories of pendulum angle φ in dependence on the excitation frequency (in the whole investigated frequency range), investigation of the influence of the mass of the fibres

At investigating the effect of the mass of the fibres on the pendulum motion results which were obtained using the "the lightest" fibres (carbon fibres; mass of one fibre is 3.846 grams), using (already mentioned) wattled steel wire parameters, using "heavier" fibres (virtual fibres of the mass ten times larger than the mass of the wattled steel wire; mass of one fibre is 177.83 grams) and using "the heaviest" fibres (virtual fibres of the mass of one fibre $1\,269$ grams) are presented in this article.

The results which were obtained using "the lightest" fibres are (almost) the same as the results which were obtained using the massless models [15]. The global extreme values of pendulum angle (with exception of using "the heaviest" fibres, where only the local extreme values are concerned) appear at excitation frequency 5 Hz irrespective of the mass of the fibres (see Fig. 9) in the monitored interval of the excitation frequencies (i.e. up to 200 Hz). The lower the mass of the fibres the higher the extreme values of the pendulum angle (see Figs. 10 and 11) at this frequency. When using "heavier" fibres pendulum angle increases at higher excitation frequencies (above 170 Hz) (see Fig. 10). When using "the heaviest" fibres pendulum angle achieves considerable local extreme value at excitation frequency 40 Hz and high extreme values of pendulum angle appear at the excitation frequency up to 130 Hz [15] — see Fig. 10.

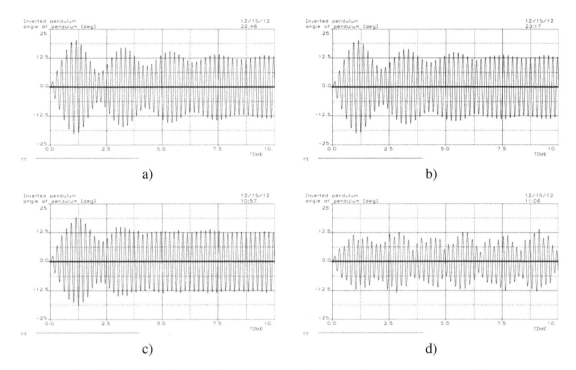

a)

b)

c)

d)

Fig. 11. Time history of pendulum angle φ, excitation frequency $f = 5$ Hz, a) carbon fibres, b) wattled steel wires, c) "heavier" fibres, d) "the heaviest" fibres, investigation of the influence of the mass of the fibres

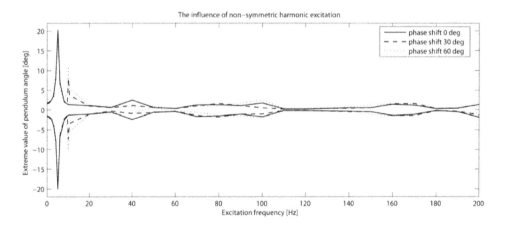

Fig. 12. Extreme values of time histories of pendulum angle φ in dependence on the excitation frequency, investigation of the influence of the phase shift in the case of non-symmetric harmonic excitation of fibres

The phase shift of in the case of non-symmetric harmonic excitation of the fibres does not manifest itself too much in the change of the character of pendulum vibration at the definite excitation frequency but especially at excitation frequency 10 Hz in the local extreme values of pendulum angle. The local extreme value of pendulum angle at this excitation frequency increases with the increasing phase shift — see Fig. 12.

5. Conclusion

The approach to the fibre modelling based on the lumped point-mass representations was utilized for the investigation of the influence of the crucial parameters of the system of the inverted pendulum driven by two fibres attached to a frame on its motion. Harmonic excitation of the fibres was considered.

Based on the obtained results it is evident that the pendulum motion is mostly influenced (besides the excitation frequency of the moving fibres) by the fibres preload [13] and by the amplitude of the harmonic kinematic excitation of fibres [14]. At the change of these parameters an unstable behaviour of the studied system was detected. Changes in other investigated parameters of this system — i.e. the change in the mass of the fibres [15] and the change of the phase shift in the case of non-symmetric harmonic excitation [12] — do not cause the unstable behaviour of the pendulum.

Experimental verification of the fibre dynamics within the manipulator systems and research aimed at measuring the material properties of selected fibres are considered important steps in further research.

Acknowledgements

The article has originated in the framework of solving No. P101/11/1627 project of the Czech Science Foundation entitled "Tilting Mechanisms Based on Fibre Parallel Kinematical Structure with Antibacklash Control" and institutional support for the long-time conception development of the research institution provided by Ministry of Industry and Trade of the Czech Republic.

References

[1] Chan, E. H. M., Design and Implementation of a High-Speed Cable-Based Parallel Manipulator, PhD Thesis, University of Waterloo, Waterloo, 2005.

[2] Diao, X., Ma, O., Vibration analysis of cable-driven parallel manipulators, Multibody System Dynamics 21 (4) (2009) 347–360.

[3] Freire, A., Negrão, J., Nonlinear Dynamics of Highly Flexible Partially Collapsed Structures, Proceedings of the III European Conference on Computational Mechanics, Solids, Structures and Coupled Problems in Engineering, Lisbon, Springer, 2006, CD-ROM.

[4] Gerstmayr, J., Sugiyama, H., Mikkola, A., Developments and Future Outlook of the Absolute Nodal Coordinate Formulation, Proceedings of the 2nd Joint International Conference on Multibody System Dynamics, Stuttgart, University of Stuttgart, Institute of Engineering and Computational Mechanics, 2012, USB flash drive.

[5] Hajžman, M., Polach, P., Modelling of Cables for Application in Cable-Based Manipulators Design, Proceedings of the ECCOMAS Thematic Conference Multibody Dynamics 2011, Université catholique de Louvain, Brussels, 2011, USB flash drive.

[6] Hajžman, M., Polach, P., Simple application of nonlinear formulation in cable mechanics, Proceedings of 10th International Conference Dynamics of Rigid and Deformable Bodies 2012, Ústí nad Labem, Univerzita J. E. Purkyně v Ústí nad Labem, 2012, CD-ROM.

[7] Kamman, J. W., Huston, R. L., Multibody Dynamics Modeling of Variable Length Cable Systems, Multibody System Dynamic 5 (3) (2001) 211–221.

[8] Liu, Ch., Tian, Q., Hu, H., New spatial curved beam and cylindrical shell elements of gradient-deficient Absolute Nodal Coordinate Formulation, Nonlinear Dynamics 70 (3) (2012) 1 903–1 918.

[9] Polach, P., Hajžman, M., Approaches to the Modelling of Inverted Pendulum Attached Using of Fibres, Proceedings of the 4th International Conference on Modelling of Mechanical and Mechatronic Systems 2011, Herľany, Technical University of Košice, 2011, CD-ROM.

[10] Polach, P., Hajžman, M., Absolute nodal coordinate formulation in dynamics of machines with cables, Proceedings of the 27th Conference with International Participation Computational Mechanics 2011, Plzeň, University of West Bohemia in Pilsen, 2011.

[11] Polach, P., Hajžman, M., Investigation of dynamic behaviour of inverted pendulum attached using of fibres, Proceedings of the 11th Conference on Dynamical Systems — Theory and Applications, Nonlinear Dynamics and Control, Łódź, Department of Automatics and Biomechanics, Technical University of Łódź, 2011, pp. 403–408.

[12] Polach, P., Hajžman, M., Investigation of dynamic behaviour of inverted pendulum attached using fibres at non-symmetric harmonic excitation, Book of abstracts of the EUROMECH Colloquium 524 Multibody system modelling, control and simulation for engineering design, Enschede, University of Twente, 2012, pp. 42–43.

[13] Polach, P., Hajžman, M., Effect of Fibre Preload on the Dynamics of an Inverted Pendulum Driven by Fibres, Proceedings of the 2nd Joint International Conference on Multibody System Dynamics, Stuttgart, University of Stuttgart, Institute of Engineering and Computational Mechanics, 2012, USB flash drive.

[14] Polach, P., Hajžman, M., Influence of the excitation amplitude on the dynamic behaviour of an inverted pendulum driven by fibres, Procedia Engineering 48 (2012) 568–577.

[15] Polach, P., Hajžman, M., Šika, Z., Mrštík, J., Svatoš, P., Effects of fibre mass on the dynamics of an inverted pendulum driven by cables, Proceedings of National Colloquium with International Participation Dynamics of Machines 2012, Prague, Institute of Thermomechanics Academy of Sciences of the Czech Republic, 2012, pp. 127–134.

[16] Polach, P., Hajžman, M., Tuček, O., Validation of the point-mass modelling approach for fibres in the inverted pendulum model, Proceedings of the 18th International Conference Engineering Mechanics 2012, Svratka, Institute of Theoretical and Applied Mechanics Academy of Sciences of the Czech Republic, 2012, CD-ROM.

[17] Polach, P., Hajžman, M., Tuček, O., Computational analysis of dynamic behaviour of inverted pendulum attached using fibres, Differential Equations and Dynamical Systems 21 (1–2) (2013) 71–81.

[18] Shabana, A. A., Flexible Multibody Dynamics: Review of Past and Recent Developments, Multibody System Dynamics 1 (2) (1997) 189–222.

[19] Smrž, M., Valášek, M., New Cable Manipulators, Proceedings of the National Conference with International Participation Engineering Mechanics 2009, Svratka, Institute of Theoretical and Applied Mechanics Academy of Sciences of the Czech Republic, 2009, CD-ROM.

[20] Valášek, M., Karásek, M., HexaSphere with Cable Actuation, Recent Advances in Mechatronics: 2008–2009, Springer-Verlag, Berlin, 2009, pp. 239–244.

[21] Wei, M. H., Xiao, Y. Q., Liu, H. T., Bifurcation and chaos of a cable-beam coupled system under simultaneous internal and external resonances, Nonlinear Dynamics 67 (3) (2012) 1969–1984.

[22] Wittbrodt, E., Adamiec-Wójcik, I., Wojciech, S., Dynamics of Flexible Multibody Systems, Rigid Finite Element Method, Springer, Berlin, 2006.

[23] Zi, B., Duan, B. Y., Du, J. L., Bao, H., Dynamic modeling and active control of a cable-suspended parallel robot, Mechatronics 18 (1) (2008) 1–12.

[24] http://hexasphere.webnode.cz

8

Linearization of friction effects in vibration of two rotating blades

M. Byrtus[a,*], M. Hajžman[a], V. Zeman[a]

[a] *Faculty of Applied Sciences, University of West Bohemia, Univerzitní 22, 306 14 Plzeň, Czech Republic*

Abstract

This paper is aimed at modelling of friction effects in blade shrouding which are realized by means of friction elements placed between blades. In order to develop a methodology of modelling, two blades with one friction element in between are considered only. Flexible blades fixed to a rotating disc are discretized by FEM using 1D Rayleigh beam elements derived in rotating space as well as the friction element modelled as a rigid body. The blades and the friction element are connected through two concurrent friction planes, where the friction forces arise on the basis of centrifugal force acting on the friction element. The linearization of friction is performed using the harmonic balance method to determine equivalent damping coefficients in dependence on the amplitudes of relative slip motion between the blades and the friction element. The methodology is applied to a model of two real blades and will be extended for the whole bladed disc with shrouding.

Keywords: bladed disc, friction, harmonic balance method, damping

1. Introduction

Blades are the common and the most important elements in turbine design. With the increase of an energy consumption turbines are still innovated and the power of developed turbines is growing. On the other hand it brings the greater complexity of newly produced energy systems and higher requirements on blades strength and fatigue. Even if a machine is properly designed with respect to excitation frequencies and turbine eigenfrequencies, some excitation sources cannot be included in preliminary developments. Therefore the blades should be designed in such a way that they can absorb vibrations caused by unexpected or unusual excitation. Mathematical and computational models of blades and their systems are suitable tools for the investigation of their dynamical properties and for their optimization.

One of the most usual approaches to the suppression of undesirable blade vibrations is the employment of various friction effects. Detailed investigation of influences of friction on dynamical response of a simplified mechanical system represented by a beam can be found in [4]. Mainly the microslip phenomenon is discussed. Another method, which is analytical one and is connected with non-spherical geometries, is developed in [1]. Many publications deal with the friction induced by means of underplatform (wedge) dampers. A method for the calculation of static balance supposing an in-plane motion of the wedge dampers is developed in [6]. An analytical approach is described in [2] and comparison of numerical simulation results with the results obtained by linearization is shown in [5]. The harmonic balance method for the evaluation of friction effects in blade dynamics represented by a very simple discrete mechanical system is discussed in [3]. Also experimental methods for the evaluation of friction significance

---

*Corresponding author. e-mail: mbyrtus@kme.zcu.cz.

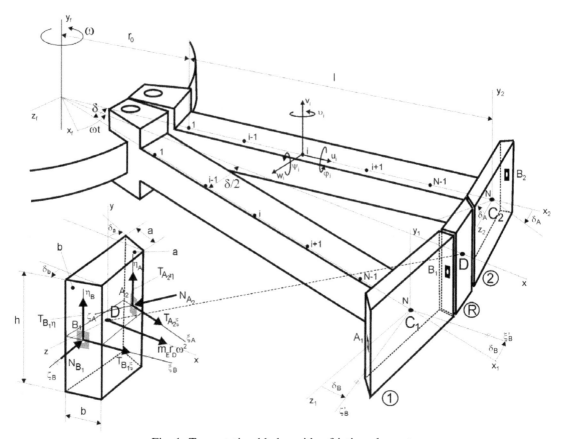

Fig. 1. Two rotating blades with a friction element

in the problems of blade vibrations are very important. Some comparison of experimental and theoretical analysis is shown in [15], pure experimental results are described in [14] and the influences of temperature are experimentally investigated in [13].

This paper deals with the modelling and dynamical analysis of a spatial system of two adjacent flexible blades with rigid shroud and one friction element, which is placed in shroud. The dynamic analysis approach is based on the harmonic balance method to replace the nonlinear friction forces with linear viscous forces determined by equivalent damping coefficients.

2. Modelling of two blades with friction coupling in rotating space

Let us consider a rigid disc with flexible blades which rotates with constant angular velocity ω in a fixed space x_f, y_f, z_f. The blade foots are fixed to the disc and every two adjacent blades are connected by means of a friction element which is wedged between the blade shrouds (see Fig. 1). As the blades rotate, the centrifugal force pushes the element towards contact surfaces a and b of the adjacent blade shroud and normal forces at the contact patches increase. Consequently, larger friction forces act on shroud in case of slip motion between the shroud and the friction element.

2.1. Rigid body

First, let us derive equations of motion of a rigid body in coordinate space x, y, z rotating with constant angular velocity ω around fixed axis y_f. The position of the body is described in rotating coordinate space x, y, z by three displacements u, v, w of the gravity centre C and

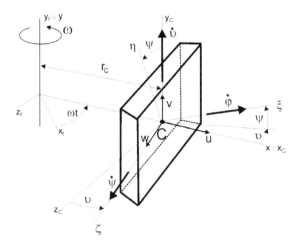

Fig. 2. Rigid body in rotating system space

three small Euler's angles φ, ϑ, ψ (see Fig. 2). Then we can formulate the kinetic energy of the body as

$$E_k = \frac{1}{2}m\boldsymbol{v}^T\boldsymbol{v} + \frac{1}{2}\boldsymbol{\omega}^T\boldsymbol{I}\boldsymbol{\omega}, \tag{1}$$

where m is mass of the body, $\boldsymbol{I}$ denotes inertia matrix of the body with respect to the coordinate system ξ, η, ζ which is fixed with the body. Vectors

$$\boldsymbol{v} = [\dot{u} + w\omega, \dot{v}, \dot{w} - (r_C + u)\omega]^T \quad \text{and} \quad \boldsymbol{\omega} = [\dot{\varphi} + \omega\psi, \omega + \dot{\vartheta}, \dot{\psi}]^T \tag{2}$$

define the velocity of the gravity centre and approximated resulting angular velocity of the body. Using the Lagrange's equation we can derive conservative model of the body in the rotating system in matrix form

$$\boldsymbol{M}_R\ddot{\boldsymbol{q}}_R + \omega\boldsymbol{G}_R\dot{\boldsymbol{q}}_R - \omega^2\boldsymbol{K}_{d,R}\boldsymbol{q}_R = \boldsymbol{f}_{\omega,R}, \tag{3}$$

where $\boldsymbol{q}_R$ is the vector of the rigid body (subscript R) generalized coordinates $\boldsymbol{q}_R = [u, v, w, \varphi, \vartheta, \psi]^T$. Matrix $\boldsymbol{M}_R$ is mass matrix, $\boldsymbol{G}_R$ is skew symmetrical matrix of gyroscopic effects and $\boldsymbol{K}_{d,R}$ constitutes matrix of softening under rotation. Force vector $\boldsymbol{f}_{\omega,R}$ expresses effects of centrifugal forces.

2.2. Single blade with shroud

Further, we will deal with modelling of a single blade using one dimensional beam finite elements in rotating system. Let us recall the mathematical model of a blade with shroud in rotating space. Detailed description and matrix derivation can be found in [7, 8, 17].

The mere blade is divided into $N - 1$ beam finite elements using N nodal points. Let us suppose, the blade foot is fixed to the rotating rigid disc at nodal point "1" (see Fig. 1). The shroud, which is the blading equipped with, is supposed to be a rigid body whose center of gravity is fixed to the last nodal point N placed at the free end of the blade (points C_1 or C_2 in the Fig. 1). Mathematical model of a decoupled rotating blade (subscript B) with shroud can be written in the matrix form [7, 17]

$$\boldsymbol{M}_B\ddot{\boldsymbol{q}}_B + (\omega\boldsymbol{G}_B + \boldsymbol{B}_B)\dot{\boldsymbol{q}}_B + (\boldsymbol{K}_{s,B} - \omega^2\boldsymbol{K}_{d,B} + \omega^2\boldsymbol{K}_{\omega,B})\boldsymbol{q}_B = \boldsymbol{f}_{\omega,B}, \tag{4}$$

where M_B, B_B, $K_{s,B}$ are symmetrical mass, material damping and stiffness matrices, respectively. Matrix G_B is skew-symmetrical and expresses gyroscopic effects. These matrices are assembled in local configuration space of the blade rotating around axis y_f with longitudinal axis x_j ($j = 1, 2$) defined by vector of generalized coordinates $q_B \in \mathbb{R}^{6N}$ having following structure

$$q_B = [\dots u_i, v_i, w_i, \varphi_i, \vartheta_i, \psi_i, \dots]^T, \quad i = 1, \dots, N, \tag{5}$$

where u_i, v_i, w_i are translational and $\varphi_i, \vartheta_i, \psi_i$ are rotational displacements of the blade in node i (see Fig. 1). Matrix $K_{d,B}$ is matrix of softening under rotation and $K_{\omega,B}$ is matrix of bending stiffening that expresses the influence of resistance in bending produced by centrifugal forces acting on the blade. The force vector $f_{\omega,B}$ describes centrifugal forces acting on blade elements at their nodes.

The model of the shroud is included in matrices M_B, G_B and $K_{d,B}$ where mass, gyroscopic and stiffness matrices of the shroud presented in (3) are added on positions corresponding to coordinates of the nodal point N. Similarly, vector $f_{\omega,B}$ of centrifugal forces is modified and the centrifugal force acting on the shroud modelled as a rigid body is added on position corresponding to the mentioned nodal coordinates.

2.3. Model of blade couple with friction element

Here, the model of two adjacent blades interconnected by a friction element will be introduced (Fig. 1). Based on previous sections, corresponding mathematical model can be written in generalized rotating coordinate system defined by vector $q = [q_1^T, q_R^T, q_2^T]^T \in \mathbb{R}^{6(N_1+1+N_2)}$ in following form

$$M\ddot{q} + (\omega G + B + B_C)\dot{q} + (K_s - \omega^2 K_d + \omega^2 K_\omega + K_C)q + h(\dot{q}) = f_\omega. \tag{6}$$

Mass, damping and stiffness matrices are arranged using models of the blades and the rigid body. Matrices in (6) are block diagonal and have this structure

$$\begin{aligned}
M &= \operatorname{diag}\left(M_1, M_R, M_2\right), \quad G = \operatorname{diag}\left(G_1, G_R, G_2\right), \\
B &= \operatorname{diag}\left(B_1, 0, B_2\right), \quad K_s = \operatorname{diag}\left(K_{s,1}, 0, K_{s,2}\right), \\
K_d &= \operatorname{diag}\left(K_{d,1}, K_{d,R}, K_{d,2}\right), \quad K_\omega = \operatorname{diag}\left(K_{\omega,1}, 0, K_{\omega,2}\right),
\end{aligned} \tag{7}$$

where indices 1 and 2 designate the first and the second blade, respectively. The index R corresponds to the friction element, which is modeled as a rigid body. The vector of excitation has the form $f_\omega = [f_{\omega,1}^T, f_{\omega,R}^T, f_{\omega,2}^T]^T$. The mathematical model (6) includes moreover coupling stiffness matrix K_C and the vector $h(\dot{q})$, which express stiffness effects and nonlinear friction forces in friction couplings between the shroud of blade 1 and 2 and the friction element R, respectively. Damping matrix proportional to contact stiffness matrix $B_C = \beta_C K_C$ comprises the influence of contact damping in contact surfaces.

2.3.1. Coupling stiffness determination

Let us deal with force effects arising in contact patches between the shroud and the friction element. The friction element has two contact surfaces a and b. At geometrical centers B_1 and A_2 of the surfaces, coordinate systems ξ_B, η_B, ζ_B and ξ_A, η_A, ζ_A are placed in such a way, that planes $\xi_B\eta_B$ and $\xi_A\eta_A$ coincide with surfaces a and b, respectively, and axes ζ_B and ζ_A are perpendicular to them (see Fig. 1, left bottom). In these coordinate systems, the forces acting

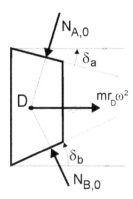

Fig. 3. Forces acting on friction element

on the friction element at point B_1 (A_2) can be expressed using normal force N_{B_1} (N_{A_2}) and friction forces $T_{B_1\xi}$ and $T_{B_1\eta}$ ($T_{A_2\xi}$ and $T_{A_2\eta}$). The resultant normal forces can be written as

$$N_{B_1} = N_{B,0} - k_b \left(\boldsymbol{\xi}_{B,C_1}^T \boldsymbol{q}_{C_1} - \boldsymbol{\xi}_{B,D}^T \boldsymbol{q}_D \right), \quad N_{A_2} = N_{A,0} + k_a \left(\boldsymbol{\xi}_{A,C_2}^T \boldsymbol{q}_{C_2} - \boldsymbol{\xi}_{A,D}^T \boldsymbol{q}_D \right), \quad (8)$$

where $N_{B,0}$ and $N_{A,0}$ are magnitudes of normal forces resulting from equilibrium conditions of non-vibrating friction element

$$N_{A,0} = mr_D\omega^2 \frac{\cos \delta_b}{\sin(\delta_a + \delta_b)}, \quad N_{B,0} = mr_D\omega^2 \frac{\cos \delta_a}{\sin(\delta_a + \delta_b)}. \quad (9)$$

Angles δ_a and δ_b describe friction element skewing (see Fig. 3). Parameters k_b and k_a express translational contact stiffnesses of contact patches and their linearized magnitudes are expressed according to

$$k_a = \frac{N_{A,0}}{\gamma_a} \cdot 10^6 [\text{N/m}], \quad k_b = \frac{N_{B,0}}{\gamma_b} \cdot 10^6 [\text{N/m}]. \quad (10)$$

Symbols γ_a and γ_b designate surface contact deformation in micrometers and are defined as follows [12]

$$\gamma_a = c\sigma_a^p, \quad \gamma_b = c\sigma_b^p, \quad (11)$$

where c is contact deformation coefficient, p is contact power index and σ_a, σ_b express average contact pressure. According to contact pressure definition, in case of neglecting of vibration influence it holds

$$\sigma_a = \frac{N_{A,0}}{A_{ef,a}}, \quad \sigma_b = \frac{N_{B,0}}{A_{ef,b}}, \quad (12)$$

where $A_{ef,a} = h_{ef}a_{ef}$ and $A_{ef,b} = h_{ef}b_{ef}$ designate supposed effective area of the corresponding contact surface, whose size can be estimated based on its experimentally gained wearing [10]. Vectors $\boldsymbol{\xi}_{XY} \in \mathbb{R}^{6,1}$ are geometric transformation vectors which transform the blade displacements in nodal point Y described by vector $\boldsymbol{q}_Y$ to a normal displacement of the contact point X at the contact surface. The terms in brackets in (8) express relative normal contact deflection between the shroud and the friction element with respect to the contact surface.

Since the blade shrouds and the friction element can rotate about their gravity center axes, it is necessary to include moreover the influence of torque in the contact surfaces. Under the assumption of identical contact stiffness in whole contact areas, the resultant torque can be described by so called rotational contact stiffnesses of contact surfaces which are defined using the translational contact stiffnesses in following way

$$k_{xx}^{(a)} = \frac{k_a}{12} \cos^2 \delta_a \, h_{ef}^2, \quad k_{xx}^{(b)} = \frac{k_b}{12} \cos^2 \delta_b \, h_{ef}^2, \quad k_{yy}^{(a)} = \frac{k_a}{12} a_{ef}^2, \quad k_{yy}^{(b)} = \frac{k_b}{12} b_{ef}^2,$$
$$k_{zz}^{(a)} = \frac{k_a}{12} \sin^2 \delta_a \, h_{ef}^2, \quad k_{zz}^{(b)} = \frac{k_b}{12} \sin^2 \delta_b \, h_{ef}^2. \tag{13}$$

Used quantities k_a and k_b are defined above as well as the meaning of angles δ_a and δ_b. The stiffnesses $k_{ax}^{(a)}$ and $k_{ax}^{(b)}$ of axial mounting of the friction element in the shroud of both blades in direction of axis y depend on the structure design. These stiffnesses influence the blade vibration very few and hinder friction element falling out. They are invariant with respect to all used coordinate systems because the axial mounting is parallel with z axes which are mutually parallel too.

Based on the above mentioned assumptions, we can express vectors of conservative forces describing mutual acting of the friction element and the blade shrouds. Because of their assumed linearity, the force vectors can be substituted by the coupling stiffness matrix presented in (6) which can be written as a sum of

$$\boldsymbol{K}_C = \boldsymbol{K}_C^{(t)} + \boldsymbol{K}_C^{(r)} + \boldsymbol{K}_C^{(ax)}, \tag{14}$$

where matrices on the right hand side describe influence of translational, rotational and axial coupling stiffnesses, respectively.

2.3.2. Friction effects determination

Now, let us deal with the nonconservative part of coupling forces. The friction forces acting on the friction element concentrated into central contact points B_1 and A_2 are nonlinear and can be expressed as

$$\overrightarrow{T}_{B_1} = f_b N_{B_1} \frac{\overrightarrow{v}_{s,B_1}}{|\overrightarrow{v}_{s,B_1}|}, \quad \overrightarrow{T}_{A_2} = f_a N_{A_2} \frac{\overrightarrow{v}_{s,A_2}}{|\overrightarrow{v}_{s,A_2}|}, \tag{15}$$

where f_b (f_a) is the friction coefficient of friction surface b (a) and $\overrightarrow{v}_{s,B_1}$ $(\overrightarrow{v}_{s,A_2})$ is slip velocity of blade shroud "1" ("2") with respect to friction element in point B_1 (A_1) expressed in $\xi_B \eta_B$ $(\xi_A \eta_A)$ plane. The friction forces acting on the blade shroud have opposite direction.

To linearize the nonlinear friction forces (15) included in (6), the harmonic balance method is used. There are many linearization techniques employed in nonlinear system investigation [9]. It depends on which kind of parameters one needs to investigate and on the excitation included in the system. The aim of the linearization technique is to replace the original nonlinear system with a linear one. The method chosen here is based on following assumptions:

1. Both nonlinear friction torques and forces acting on a friction element interact mutually very weak, therefore equivalent damping coefficients can be considered independently.

2. The slip motion of friction surfaces can be simply considered as one degree of freedom motion in the direction of the slip.

3. Excitation is supposed to be a periodic function as well as the steady-state response.

4. The friction and excitation forces are expandable into a Fourier series.

Based on this, the term for determination of equivalent damping coefficient for k-th harmonic component with angular frequency ω_k [16] can be derived in following form

$$b_e(a_k, \omega_k) = \frac{4T}{\pi a_k \omega_k}, \qquad (16)$$

where T is the magnitude of friction force, a_k is the amplitude of steady slip motion and ω_k is excitation angular frequency. According to known experimental observations, the term (16) does not fit real, measured amplitudes of slip motion. In [11], a modification of (16) is suggested

$$b_e(a_k, \omega_k) = \frac{4T}{\pi (a_k \omega_k)^{1.112}}. \qquad (17)$$

Physically, the term $a_k \omega_k$ presents the amplitude of corresponding harmonic component of the slip velocity. Using the modified equivalent damping coefficient (17), each harmonic component of nonlinear friction forces (15) can be linearized and expressed by the coefficient of equivalent viscous translational and rotational damping (e.g. for the point B_1)

$$b_e^{(t)}(a_{B,k}^{(t)}, \omega_k) = \frac{4T}{\pi (a_{B,k}^{(t)} \omega_k)^{1.112}}, \quad b_e^{(r)}(a_{B,k}^{(r)}, \omega_k) = \frac{4M}{\pi (a_{B,k}^{(r)} \omega_k)^{1.112}}. \qquad (18)$$

Variables $a_{B,k}^{(t)}$ and $a_{B,k}^{(r)}$ constitute translational and rotational slip amplitudes excited by k-th harmonic component, respectively. The linearized friction forces in (15) and friction torque M_{B_1} acting on the friction element can be then rewritten using (16) into (again for the point B_1)

$$\overrightarrow{T}_{B_1} = \frac{4T}{\pi (a_B^{(t)} \omega_0)^{1.112}} \overrightarrow{v}_{s,B_1}, \quad M_{B_1} = \frac{4M}{\pi (a_B^{(r)} \omega_0)^{1.112}} \dot{\psi}_B, \qquad (19)$$

where magnitude of friction force reads $T = f_b N_{B,0}$ and providing the same contact pressure on a circular surface with radius r_{ef}, magnitude M of friction torque is $M = \frac{2}{3} f_b r_{ef} N_{B,0}$. Variable $\dot{\psi}_B$ designates relative rotational slip velocity of the blade "1" with respect to friction element in $\xi_B \eta_B$ plane, i.e. about ζ_B axis (see Fig. 2).

For illustration, ξ-component of friction force expressed in coordinate system of the friction element states as

$$T_{B_1 \xi} = b(a_{B\varphi}) c_{B\xi}, \qquad (20)$$

where $a_{B\xi}$ is the slip amplitude in ξ-direction and $c_{B\xi}$ represents ξ-component of slip velocity and can be expressed as

$$c_{B\xi} = \boldsymbol{\tau}_{B,C_1}^T \dot{\boldsymbol{q}}_{C_1} - \boldsymbol{\tau}_{B,D}^T \dot{\boldsymbol{q}}_D. \qquad (21)$$

Quantities $\boldsymbol{\tau}_{B,C_1}^T$ and $\boldsymbol{\tau}_{B,D}^T$ are shown in (29).

Finally, the nonlinear mathematical model (6) can be equivalently replaced by linearized one for each excitation harmonic component with frequency ω_0

$$\boldsymbol{M}\ddot{\boldsymbol{q}} + (\omega \boldsymbol{G} + \boldsymbol{B} + \boldsymbol{B}_C + \boldsymbol{B}_e(\boldsymbol{a}, \omega_0))\dot{\boldsymbol{q}} + (\boldsymbol{K}_s - \omega^2 \boldsymbol{K}_d + \omega^2 \boldsymbol{K}_\omega + \boldsymbol{K}_C)\boldsymbol{q} = \boldsymbol{f}_\omega + \boldsymbol{f}(\omega_0)e^{i\omega_0 t}. \quad (22)$$

Friction torques and forces are represented by equivalent damping matrix $\boldsymbol{B}_e(\boldsymbol{a}, \omega_0)$, where $\boldsymbol{a} = [a_A^{(t)}, a_A^{(r)}, a_B^{(t)}, a_B^{(r)}]^T$ is a vector containing steady slip amplitudes. Vector $\boldsymbol{f}(\omega_0)$ of complex amplitudes represents external harmonic excitation with frequency equal to ω_0.

2.3.3. *Equivalent damping calculation for steady-state response*

Even if the terms in (18) hold, they depend on slip amplitude of the examined motion. For this purpose, it is necessary to perform any estimation of this amplitude. To approach the real slip amplitude as close as possible we use the linearized model neglecting the equivalent damping matrix. Further, let us point out, that the force vector $\boldsymbol{f}_\omega$ in (22) includes constant centrifugal forces in the rotating coordinate system and therefore corresponding response is a constant vector of displacements. This vector has also no influence on the equivalent damping coefficients. In the next, the influence of centrifugal forces can be neglected and the equivalent damping coefficients are calculated for harmonic excitation. To gain the global response to harmonic and centrifugal (static) excitation the superposition law for linear systems can be advantageously used.

Let us find the steady-state solution of (22) in rotating system space in complex domain in this form

$$\boldsymbol{q}(t) = \widetilde{\boldsymbol{q}}(\omega_0)e^{\mathrm{i}\omega_0 t}, \tag{23}$$

where $\widetilde{\boldsymbol{q}}$ is a vector of complex amplitudes of displacements and $\mathrm{i} = \sqrt{-1}$. To determine the complex amplitude, let us put (23) in (22) for $\boldsymbol{f}_\omega = 0$, $\boldsymbol{B}_e = 0$ and we obtain

$$\widetilde{\boldsymbol{q}}(\omega_0) = \left[-\omega_0^2 \boldsymbol{M} + \mathrm{i}\omega_0(\omega\boldsymbol{G} + \boldsymbol{B} + \boldsymbol{B}_C) + (\boldsymbol{K}_s - \omega^2\boldsymbol{K}_d + \omega^2\boldsymbol{K}_\omega + \boldsymbol{K}_C) \right]^{-1} \boldsymbol{f}(\omega_0), \tag{24}$$

where $\boldsymbol{f}(\omega_0)$ is vector of complex amplitudes of harmonic excitation. Based on the vector of complex amplitudes of displacements, the complex amplitudes of slip motion between contact surfaces can be determined. Let us focus on contact point B_1. The complex translational slip amplitude in $\xi_B\eta_B$ plane can be defined as

$$\widetilde{a}_B = \widetilde{a}_{B\xi} + \mathrm{i}\,\widetilde{a}_{B\eta} = \overline{a}_{B\xi} + \mathrm{i}\,\overline{\overline{a}}_{B\xi} + \mathrm{i}(\overline{a}_{B\eta} + \mathrm{i}\,\overline{\overline{a}}_{B\eta}), \tag{25}$$

where $\widetilde{a}_{B\xi}$ and $\widetilde{a}_{B\eta}$ are complex slip amplitudes in the direction of ξ_B and η_B axis, respectively, $\overline{a}$ and $\overline{\overline{a}}$ denote real and complex part. The complex amplitudes are determined for contact point B_1 as follows

$$\widetilde{a}_{B\xi} = \boldsymbol{\xi}_{B,C_1}^T \boldsymbol{q}_{C_1} - \boldsymbol{\xi}_{B,D}^T \boldsymbol{q}_D, \quad \widetilde{a}_{B\eta} = \boldsymbol{\eta}_{B,C_1}^T \boldsymbol{q}_{C_1} - \boldsymbol{\eta}_{B,D}^T \boldsymbol{q}_D, \tag{26}$$

for contact point A_2

$$\widetilde{a}_{A\xi} = \boldsymbol{\xi}_{A,C_2}^T \boldsymbol{q}_{C_2} - \boldsymbol{\xi}_{A,D}^T \boldsymbol{q}_D, \quad \widetilde{a}_{A\eta} = \boldsymbol{\eta}_{A,C_2}^T \boldsymbol{q}_{C_2} - \boldsymbol{\eta}_{A,D}^T \boldsymbol{q}_D \tag{27}$$

and complex rotational slip amplitudes

$$\widetilde{a}_{A\varphi} = \boldsymbol{\tau}_{A,C_2}^T \boldsymbol{q}_{C_2}^{(r)} - \boldsymbol{\tau}_{A,D}^T \boldsymbol{q}_D^{(r)}, \quad \widetilde{a}_{B\varphi} = \boldsymbol{\tau}_{B,C_2}^T \boldsymbol{q}_{C_2}^{(r)} - \boldsymbol{\tau}_{B,D}^T \boldsymbol{q}_D. \tag{28}$$

Transformation vectors $\boldsymbol{\xi}_{X,Y}\,(\boldsymbol{\eta}_{X,Y}) \in \mathbb{R}^{6,1}$ transform the blade displacements in nodal point Y defined by vector $\boldsymbol{q}_Y$ to a translational displacement in ξ (η) direction of contact point X laying in the contact surface. Vectors $\boldsymbol{\tau}_{X,Y} \in \mathbb{R}^{3,1}$ transform rotational displacements in nodal point Y described by vector $\boldsymbol{q}_Y^{(r)}$ to rotational displacements in contact point X. Transformation vectors have following structure

$$\boldsymbol{\xi}_{B,C_1}^T = [\sin \delta_B, 0, \cos \delta_B, 0, -\overline{BC}_1 \sin \delta_B, 0],$$
$$\boldsymbol{\xi}_{B,D}^T = [\sin \delta_b, 0, \cos \delta_b, 0, \overline{BD} \sin \delta_b, 0],$$
$$\boldsymbol{\eta}_{B,C_1}^T = [0, 1, 0, \overline{BC}_1, 0, 0],$$
$$\boldsymbol{\eta}_{B,D}^T = [0, 1, 0, -\overline{BD}, 0, 0],$$
$$\boldsymbol{\xi}_{A,C_2}^T = [-\sin \delta_A, 0, \cos \delta_A, 0, -\overline{AC}_2 \sin \delta_A, 0],$$
$$\boldsymbol{\xi}_{A,D}^T = [-\sin \delta_a, 0, \cos \delta_a, 0, \overline{AD} \sin \delta_a, 0], \tag{29}$$
$$\boldsymbol{\eta}_{A,C_2}^T = [0, 1, 0, -\overline{AC}_2, 0, 0],$$
$$\boldsymbol{\eta}_{A,D}^T = [0, 1, 0, \overline{AD}, 0, 0],$$
$$\boldsymbol{\tau}_{A,C_2}^T = [\cos \delta_A, 0, \sin \delta_A, 0, \overline{AC}_2 \cos \delta_A, 0],$$
$$\boldsymbol{\tau}_{A,D}^T = [\cos \delta_a, 0, \sin \delta_a, 0, -\overline{AD} \cos \delta_a, 0],$$
$$\boldsymbol{\tau}_{B,C_1}^T = [\cos \delta_B, 0, -\sin \delta_B, 0, -\overline{BC}_1 \cos \delta_B, 0],$$
$$\boldsymbol{\tau}_{B,D}^T = [\cos \delta_b, 0, -\sin \delta_b, 0, \overline{BD} \cos \delta_b, 0].$$

The subscripts of the contact point designation have been left out according to relations (25)–(29).

Using the relations (25)–(28), equivalent damping coefficients (18) can be evaluated for both contact surfaces and for each harmonic component of excitation. Instead of the amplitude of corresponding slip motion, the absolute value of complex amplitudes (25) has to be used. Expressing the vector of slip amplitudes $\boldsymbol{a} = [\,|\tilde{a}_A^{(t)}|, |\tilde{a}_A^{(r)}|, |\tilde{a}_B^{(t)}|, |\tilde{a}_B^{(r)}|\,]^T$ we can determine the equivalent damping matrix in (22) and evaluate the vector of complex amplitudes of displacements taking into account the linearized nonlinear friction effects.

3. Application

The methodology of the modelling presented above is used for dynamic analysis of a real blade couple (see Fig. 4). Although the real blade packet consists of five blades connected with four friction elements, the methodology is proved on one blade couple connected with one friction body. The blades are fixed to a rigid disc rotating with constant angular velocity. Detailed geometrical description of the blades was gained from [10]. Based on the derived methodology, in-house software for computational blade modelling was developed. Using this software, each blade was divided by six nodal points into five finite beam elements. Final computational model has 78 DOF (two blades and one friction body). Basic geometrical parameters (see Fig. 1) used during numerical calculations are: $b = 0.006$ m, $h = 0.02$ m, $l = 0.21$ m, $r_D = 0.4655$ m, $\delta_a = 20°$, $\delta_b = 0°$, weight of friction element $m = 0.008\,6$ kg. Parameters used for contact stiffnesses calculations $c = 3$, $p = 0.5$. Dimensions of effective are contact surface are supposed to be $h_{a\,ef} = h_{b\,ef} = 0.016$ m, $a_{ef} = 0.005\,1$ m, $b_{ef} = 0.004\,8$ m and the friction coefficient f is varied from 0.1 to 0.3.

3.1. Modal analysis

The derived linearized model (22) is used as a first approximation of the nonlinear behaviour of the blade packet. Performing the modal analysis, we can see the influence of friction damping on the spectrum of natural frequencies.

Fig. 4. Blade packet with friction elements

Table 1. Chosen natural frequencies of blade couple — conservative model

| ν | Natural frequency [Hz] | | | Mode shape |
	0 rpm	2 000 rpm	3 000 rpm	
1	133.8	136.9	140.7	blade bending in xy plane
2	153.7	198.2	197.9	blade bending in xy plane, FE displacement
3	158.2	267.5	268.2	blade bending in xz plane
4	178.3	540.4	561.4	dominant blade torsion
5	266.8	955.8	959.7	blade bending in in xy plane
6	356.4	1 012.5	1 020.2	blade torsion
7	449.2	1 255.5	1 270.8	blade bending in xz plane
8	724.2	1 680.4	1 693.7	blade bending in xz plane
9	958.8	1 866.3	1 868.1	blade bending in xz plane
10	959.1	2 622.2	2 716.6	blade torsion
11	1 058.4	2 729.6	2 775.2	blade bending in yz plane, FE twisting about z axis
12	1 554.2	2 881.5	3 294.2	blade torsion, FE twisting about z axis

Abbreviation FE denotes Friction Element.

3.1.1. Conservative model

To gain natural frequencies and corresponding mode shapes of the conservative model, we perform the modal analysis of the model (22) where we put $B = B_e = B_C = 0$. Because of the presence of matrices of softening and bending stiffening under rotation and of the matrix of gyroscopic effects in the model, natural frequencies depend on angular velocity $\omega = \pi n/30$ rad/s, where n designates rotational speed of the disc defined in revolutions per minute. The eigenvalues are complex conjugate with zero real part. Imaginary parts of chosen eigenvalues are presented in the Table 1 along with the description of corresponding mode shapes. As the disc

Table 2. Influence of friction on real parts of eigenvalues — nonconservative model

ν	Eigenvalues f_ν [Hz]				
	$f = 0$	$f = 0.1$	$f = 0.2$	$f = 0.3$	$f = 0.4$
1	$-0.6 + 136.9i$	-0.002	-0.001	$-0.000\,7$	$-0.000\,5$
2	$-1.2 + 198.8i$	-0.1	-0.05	-0.02	-0.01
3	$-2.2 + 267.5i$	$-2.7 + 188i$	$-4.3 + 188.2i$	-0.1	-0.1
4	$-9.1 + 540.2i$	$-2.4 + 267.5i$	$-2.5 + 267.5i$	$-5.9 + 188.5i$	$-7.5 + 188.9i$
5	$-28.5 + 955.4i$	$-21.4 + 594.2i$	$-31.4 + 595.5i$	$-2.7 + 267.5i$	$-2.8 + 267.5i$
6	$-32 + 1\,012i$	$-52 + 1\,001i$	$-72 + 1\,008i$	$-41 + 598i$	$-50.5 + 601.4i$
7	$-49 + 1\,254i$	$-52 + 1\,205i$	$-59 + 1\,207i$	$-89 + 1\,018i$	$-102 + 1\,031i$
8	$-88 + 1\,678i$	$-78 + 1\,562i$	$-77 + 1\,562i$	$-65 + 1\,209i$	$-71 + 1\,213i$
9	$-109 + 1\,863i$	$-113 + 1\,857i$	$-117 + 1\,857i$	$-77 + 1\,562i$	$-77 + 1\,562i$
10	$-215 + 2\,613i$	$-160 + 2\,246i$	$-161 + 2\,247i$	$-122 + 1\,858i$	$-126 + 1\,856i$
11	$-233 + 2\,719i$	$-326 + 2\,745i$	$-411 + 2\,764i$	$-161 + 2\,247i$	$-161 + 2\,248i$
12	$-260 + 2\,870i$	$-544 + 3\,287i$	$-545 + 3\,459i$	$-490 + 2\,805i$	$-552 + 2\,868i$

All presented eigenvalues has been calculated for 2 000 rpm and for nozzle excitation with thirty fold times higher frequency than disc speed.

rotates, stiffening effects under rotation increase and values of natural frequencies increase too, except the lowest one. Mode shapes and natural frequencies under rotation are moreover influenced by the lock-effect, which is caused by centrifugal forces acting on friction element. After locking, blade shrouds are interconnected as if they are one rigid element and between the friction areas the micro-slip motion appears only.

3.1.2. Nonconservative model

Taking into account the influence of contact and friction damping, number of complex eigenvalues vanishes and real eigenvalues appear instead. This is the desired positive effect of friction damping because corresponding mode shapes are also super-critically damped. This effect can be clearly seen in Table 2, where first twelve eigenvalues are presented for different contact friction coefficients. For example, if friction coefficient $f = 0.2$ and rotational speed is 2 000 rpm, we have 47 complex conjugate eigenvalues and 62 negative real eigenvalues. The number of complex and real eigenvalues changes slightly along with rotational speed of the disc. Further, because of the influence of matrices of softening and bending stiffening under rotation and because of the coupling friction effects, imaginary parts of eigenvalues depend not only on rotational speed but also on excitation frequency. This effect can be clearly seen in Fig. 5, where imaginary parts of first ten eigenvalues are plotted in dependence on rotational speed of the disc. The straight line corresponds to dependence of excitation with frequency thirtyfold higher than rotational frequency of the disc. This case corresponds to undermentioned considered excitation frequency. Further, it is worthy to be noticed here that the considered excitation is defined in (30) and it acts in axial direction of the disc. Therefore, imaginary parts of eigenvalues corresponding to mode shapes dominantly vibrating in axial direction are influenced only (i.e. the fourth, the sixth, the eighth and the tenth mode). As an illustration, in Fig. 6, chosen mode shapes are plotted considering damped model ($f = 0.2$) rotating with rotational velocity of 2 000 rpm.

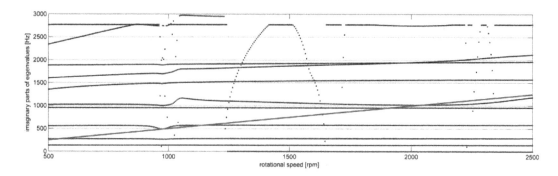

Fig. 5. Dependence of imaginary parts of eigenvalues on rotational speed of the disc and on excitation frequency

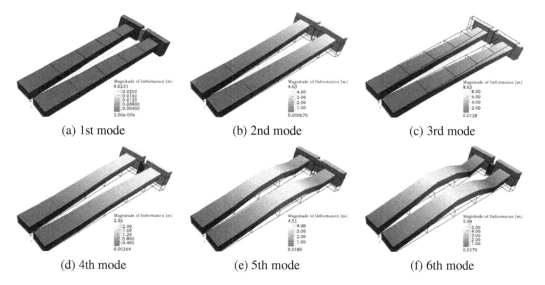

(a) 1st mode   (b) 2nd mode   (c) 3rd mode

(d) 4th mode   (e) 5th mode   (f) 6th mode

Fig. 6. Chosen mode shapes corresponding to damped model

3.2. Steady-state response to external excitation

The model derived above was used for determination of steady-state response to arbitrary harmonic component of periodic excitation with frequency corresponding to angular velocity ω of the disc. The excitation considered here simulates electromagnetic pulses acting on the blade shroud. The electromagnets are placed in the fixed nonrotating system and act on the blades at an instant of blade passage around the electromagnet (see Fig. 4). Since the blades rotate with constant angular velocity ω, the electromagnetic pulses acting on each blade are mutually delayed by $\Delta t = \frac{2\pi}{\omega n_B}$, where n_B denotes number of blades uniformly distributed around the circumference of the rotating disc. By the reason that the excitation regarding bladed disc is periodic with basic frequency equal to rotational frequency of the disc, the vector of complex amplitudes of external periodic excitation in (24) can be determined using Fourier series with the first K terms in following form

$$\boldsymbol{f}(\omega, t) = [\ldots, 0, \ldots, e^{-i\omega\Delta t}, \ldots, 0, \ldots, 1, \ldots, 0, \ldots]^T \sum_{k=1}^{K} F_k e^{ik\omega t}, \quad F_k = F_0/k, \quad (30)$$

where the two nonzero elements denote complex amplitudes of excitation with fundamental frequency ω and correspond to coordinates describing axial displacements (in direction of rotation axis) of and nodes C_1 and C_2 of both blade shrouds. The excitation of the first blade is time-shifted about Δt which designates the time between passage of baldes around the electromagnet. Based on the periodic excitation (30), the steady state response can be calculated for each harmonic component of excitation.

3.3. Steady-state calculation procedure

The calculation procedure of steady-state response of two blades with shroud can be summarized as follows:

1. Creation of the model of blade couple with friction element considering material damping only and neglecting both friction forces in contact planes ($f = 0$), i.e. determining the matrices in (7).

2. Contact stiffness and damping matrices determination according to (8)–(14) and further detail described in Appendices A and B.

3. Mathematical model creation without friction forces effects.

4. Steady-state response calculation for given harmonic excitation (24) without friction forces effects.

5. Calculation of complex slip amplitudes of steady-state harmonic motion according to (25)–(28) in contact points A and B.

6. Determination of coefficients of equivalent viscous translational and rotational damping (18) using absolute values of complex slip amplitudes in contact points A and B.

7. Composition of equivalent damping matrix based on equivalent viscous coefficients.

8. More accurate steady-state response calculation for given harmonic excitation considering the equivalent viscous friction damping.

Following above mentioned eight steps one can get steady-state response of a blade couple considering the friction effects in the blade shroud. Steps 4 to 8 can be used as an iteration process when the steady-state response from step 8 is chosen to be an initial condition for the second and the later iterations. In the next, presented results have been gained using the first iteration only because the iteration process do not converge due to the equivalent damping coefficients expression (18). The point is that taking into account the equivalent damping coefficient, the steady-state response decreases along with slip amplitudes and therefore the equivalent damping coefficient increases. Also, this approach can serve as a first approximation of the nonlinear model solution only.

As an illustration, in Fig. 7 absolute values of complex amplitudes of translational displacements of the first blade shroud are plotted in dependence on rotational speed for $k = 30$ and $F_0 = 100$ N. The friction coefficients of both friction surfaces are equal $f_a = f_b$. The considered harmonic part of excitation corresponds to nozzle frequency ($k = n_B$). Then resonant speeds $n_{\nu,k} = 60/k\mathrm{Im}\{f_\nu\}$ rpm in given speed range correspond to ν-th natural frequency excited by k-th harmonic component of excitation.

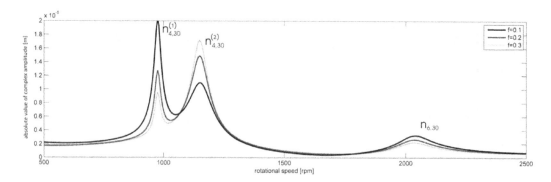

Fig. 7. Absolute values of complex amplitudes of translational displacement of the first blade shroud in direction of rotation axis for different friction coefficients

One can see three resonant peaks in Fig. 7 corresponding to the fourth and to the sixth natural frequency, $n_{4,30}^{(1)} = 980$ rpm, $n_{4,30}^{(2)} = 1\,150$ rpm and $n_{6,30} = 2\,045$ rpm. Taking into account the mode shapes plotted in Fig. 6 and the axial direction of excitation, the resonant frequencies coincide with natural frequencies corresponding to bending mode shapes in xy plane. The first two peaks correspond to the fourth frequency because of the dependence of imaginary parts of eigenvalues on rotational speed and on excitation frequency. Comparing Fig. 5 and Fig. 7 the leading line intersects the fourth frequency in two points. Changing the friction coefficient, the values of fourth frequency draw apart the leading line in the area of first resonant peak. And that is the reason why the first peak is less dominant for higher values of friction coefficient.

4. Conclusion

This paper presents a method focused on modelling of friction effects in blade shroud which are realized by means of friction elements placed in between the blade shroud. A model of two rotating blades with shroud is derived and can be easily generalized for complete bladed disc as well as the developed methodology for linearization of nonlinear friction forces acting between the blades and the friction element. The first gained results confirm that the efficiency of friction forces is dominant in resonant states. The linearization of friction forces is based on the harmonic balance method which is used for equivalent damping coefficients determination in dependence on the amplitudes of relative slip motion between blades and the friction element. According to the methodology the in-house software in MATLAB was created and tested on a model of two rotating blades with shroud.

Acknowledgements

This work was supported by the GA CR project No. 101/09/1166 "Research of dynamic behaviour and optimization of complex rotating systems with non-linear couplings and high damping materials".

References

[1] Allara, M., A model for the characterization of friction contacts in turbine blades, Journal of Sound and Vibration 320 (2009) 527–544.

[2] Borrajo, J. M., Zucca, S., Gola, M. M., Analytical formulation of the Jacobian matrix for nonlinear calculation of forced response of turbine blade assemblies with wedge friction dampers, International Journal of Non-Linear Mechanics 41 (2006) 1 118–1 127.

[3] Cha, D., Sinha, A., Statistics of responses of a mistuned and frictionally damped bladed disk assembly subjected to white noise and narrow band excitations, Probabilistic Engineering Mechanics 21 (2006) 384–396.

[4] Cigeroglu, E., An, N., Menq, C. H., A microslip friction model with normal load variation induced by normal motion, Nonlinear Dynamics 50 (2007) 609–626.

[5] Csaba, G., Forced response analysis in time and frequency domains of a tuned bladed disk with friction dampers, Journal of Sound and Vibration 214 (3) (1998) 395–412.

[6] Firrone, C. M., Zucca, S., Underplatform dampers for turbine blades: The effect of damper static balance on the blade dynamics, Mechanics Research Communications 36 (2009) 515–522.

[7] Kellner, J., Vibration of turbine blades and bladed disks, Ph.D. thesis, University of West Bohemia, Pilsen, 2009. (in Czech)

[8] Byrtus, M., Hajžman, M., Zeman, V., Dynamics of rotating systems, University of West Bohemia, Pilsen, 2010. (in Czech)

[9] Nayfeh, A. H., Mook, D. T., Nonlinear oscillations, Wiley-VCH Verlag, Weinheim, 2004.

[10] Půst, L., Veselý, J., Horáček, J., Radolfová, A., Research on friction effects in blades model, Institute of Thermomechanics, Academy of Sciences of the Czech Republic, Prague, 2008. (research report, in Czech)

[11] Půst, L., Horáček, J., Radolfová, A., Beam vibration with a friction surface on perpendicular extenders, Institute of Thermomechanics, Academy of Sciences of the Czech Republic, Prague, 2008. (research report, in Czech)

[12] Rivin, E. I., Stiffness and damping in mechanical design, Marcel Dekker, New York, 1989.

[13] Schmidt-Fellner, A., Siewert, C., Panning, L., Experimental analysis of shrouded blades with friction contact, Proceedings in Applied Mathematics and Mechanics 6 (2006) 263–264.

[14] Tokar, I. G., Zinkovskii, A. P., Matveev, V. V., On the problem of improvement of the damping ability of rotor blades of contemporary gas-turbine engines, Strength of Materials 35 (4) (2003) 368–375.

[15] Toufine, A., Barrau, J. J., Berthillier, M., Dynamics study of a structure with flexion-torsion coupling in the presence of dry friction, Nonlinear Dynamics 18 (1999) 321–337.

[16] Rao, S. S., Mechanical vibrations (Fourth Edition), Pearson Prentice Hall, 2004.

[17] Zeman, V., Kellner, J., Modelling of vibration of moving blade packet, Modelling and Optimization of Physical systems, Zeszyty naukowe 31 (2006), 155–160, Gliwice.

Appendix A — Contact stiffness

Let us summarize the determination of contact stiffnesses. The contact stiffnesses can be determined based on expression of contatct forces and contact torques transmitted by contact surfaces. As the first step, let us express the torques caused by contact forces, which are acting at friction element by supposed centric placed rectangular effective surfaces a and b, can be expressed under assumption of constant surface stiffnesses $\kappa_a = k_a/(h_{aef}a_{ef})$, $\kappa_b = k_b/(h_{bef}b_{ef})$ along the whole contact area. Using this assumption, torques of contact forces, e.g. in surface b, can be expressed in local coordinate system ξ_B, η_B, ζ_B (see Fig. 8) under assumption of relative angular displacements $\varphi_{\xi_B}, \varphi_{\eta_B}, \varphi_{\zeta_B}$ of friction element (surface b) around axes ξ_B, η_B, ζ_B with respect to the shroud as

$$M_{\xi_B}^{(b)} = 2 \int_0^{h_{bef}/2} \underbrace{\kappa_b \overbrace{b_{ef} d\eta}^{dA}}_{dN} \underbrace{\eta \varphi_{\xi_B}}_{\text{deformation}} \quad \eta = \frac{1}{12} \underbrace{\kappa_b b_{ef} h_{bef}}_{k_{\zeta_B}} h_{bef}^2 \varphi_{\xi_B} = k_{\xi_B \xi_B}^{(b)} \varphi_{\xi_B},$$

$$M_{\eta_B}^{(b)} = 2 \int_0^{b_{ef}/2} \underbrace{\kappa_b \overbrace{h_{bef} d\xi}^{dA}}_{dN} \underbrace{\xi \varphi_{\eta_B}}_{\text{deformation}} \quad \xi = \frac{1}{12} \underbrace{\kappa_b b_{ef} h_{bef}}_{k_{\zeta_B}} b_{ef}^2 \varphi_{\xi_B} = k_{\eta_B \eta_B}^{(b)} \varphi_{\eta_B}.$$

Based on the placement of the coordinate system ξ_B, η_B, ζ_B, there appears no contact force by rotational motion around the ζ_B axis which is perpendicular to the surface b. Therefore, corresponding contact torque is zero too

$$M_{\zeta_B}^{(b)} = 0.$$

Linearized rotational contact stiffnesses in surface b with respect to axes ξ_B, η_B, ζ_B of friction element result from above given terms, i.e.

$$k_{\xi_B \xi_B}^{(b)} = \frac{1}{12} k_{\zeta_B} h_{bef}^2, \quad k_{\eta_B \eta_B}^{(b)} = \frac{1}{12} k_{\zeta_B} b_{ef}^2, \quad k_{\zeta_B \zeta_B}^{(b)} = 0.$$

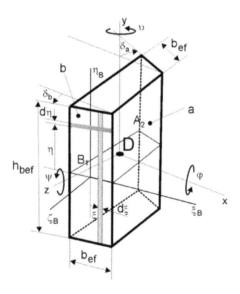

Fig. 8. Friction element detail

Derived contact stiffnesses are expressed in coordinate systems connected with contact surfaces. To express contact forces acting on friction element and blade ends, these forces has to be transformed in proper coordinate systems.

Appendix B — Contact forces and structure of contact matrices

In contact points B_1, A_2 placed between blade shrouds and friction element, contact elastic forces and torques are acting. These force effects are transformed into blade end points C_1 and

C_2 and into the center of mass D of friction element. In configuration space of generalized coordinates $q \in \mathbb{R}^{6(N_1+1+N_2)}$, where N_1 and N_2 are numbers of blade nodes, vector of elastic couplings forces can be expressed in the form

$$K_C q = \left[\dots, f_{C_1}^T, m_{C_1}^T, f_a^T + f_b^T, m_a^T + m_b^T, \dots, f_{C_2}^T, m_{C_2}^T \right]^T. \tag{31}$$

Vectors f_{C_1}, f_{C_2} and m_{C_1}, m_{C_2} represent forces and torques by which friction element acts on blades in nodes C_1 and C_2. Vectors f_a, f_b and m_a, m_b express forces and torques, respectively, by which the shrouds act on friction element in mass center D. Particular vectors have nonzero components regarding axes x_j, y_j, z_j $(j = 1, 2)$ and x, y, z.

Let us express particular forms of force and torque vectors presented in equation (31). Vectors f_{C_1} and m_{C_1} express effects of friction element on the blade in node C_1. These vectors are expressed in coordinate system x_1, y_1, z_1 where the origin is identical to point C_1. Vector f_{C_1} includes effects of normal and friction forces acting at the blade in contact point B_1 and have following form

$$f_{C_1} = \begin{bmatrix} -N_{B_1} \sin \delta_B + T_{B_1 \xi} \cos \delta_B \\ T_{B_1 \eta} \\ -N_{B_1} \cos \delta_B - T_{B_1 \xi} \sin \delta_B \end{bmatrix}.$$

Symbols $T_{B_1 \xi}$ and $T_{B_1 \eta}$ represent ξ- and η-components of friction force T_{B_1}. Further, the vector expressing the torque can be expressed as follows

$$m_{C_1} = R_{B,C_1} f_{C_1},$$

where $R_{B,C_1} \in \mathbb{R}^{3,3}$ is a matrix of vector product and has following antisymmetric structure

$$R_{B,C_1} = \begin{bmatrix} 0 & -z_{B_1} & y_{B_1} \\ z_{B_1} & 0 & 0 \\ -y_{B_1} & 0 & 0 \end{bmatrix}. \tag{32}$$

Nonzero elements in the term above represent coordinates of contact point B_1 in coordinate system x_1, y_1, z_1. Let us follow next terms in equation (31). The vector f_a represents force effect of the blade shroud acting on friction element in the surface a with the centre A_2 and it has following form

$$f_a = \begin{bmatrix} -N_{A_2} \sin \delta_a + T_{A_2 \xi} \cos \delta_a \\ T_{A_2 \eta} \\ N_{A_2} \cos \delta_a + T_{A_2 \xi} \sin \delta_a \end{bmatrix}.$$

Similarly, the vector f_b has the form

$$f_b = \begin{bmatrix} -N_{B_1} \sin \delta_b + T_{B_1 \xi} \cos \delta_b \\ T_{B_1 \eta} \\ -N_{B_1} \cos \delta_b - T_{B_1 \xi} \sin \delta_b \end{bmatrix}.$$

Corresponding torques acting on the friction element can be expressed similarly as above

$$m_a = R_{A,D} f_a, \quad m_b = R_{B,D} f_b.$$

Matrices $R_{A,D}$ and $R_{B,D}$ have the structure as shown in (32). Vectors f_{C_2} and m_{C_2} are then expressed following the same steps as above using corresponding values.

It is efficient to split their expression to translational contact deformations in dependence on translational contact stiffnesses k_a, k_b, rotational contact deformations in dependence on rotational contact stiffnesses derived in Appendix A and to axial contact deformation in dependence on axial mounting stiffnesses $k_{ax}^{(a)}$ and $k_{ax}^{(b)}$ of friction element in blade shroud. Therefore, the vector of elastic coupling forces can be written in separated form

$$\boldsymbol{K}_C \boldsymbol{q} = \left(\boldsymbol{K}_C^{(t)} + \boldsymbol{K}_C^{(r)} + \boldsymbol{K}_C^{(ax)} \right) \boldsymbol{q},$$

to which the coupling matrix (14) corresponds. Structure of these matrices is given below, where crosshatch blocks designate nonzero block matrices of order 6 as well as the dimension of vectors of generalized coordinates of corresponding nodes.

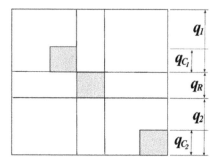

Fig. 9. Coupling stiffness matrices

9

Various methods of numerical estimation of generalized stress intensity factors of bi-material notches

J. Klusák[a,*], T. Profant[b], M. Kotoul[b]

[a] *Institute of Physics of Materials, Academy of Sciences of the Czech Republic, Žižkova 22, 616 62 Brno, Czech Republic*
[b] *Brno University of Technology, Technická 2, 616 69 Brno, Czech Republic*

Abstract

The study of bi-material notches becomes a topical problem as they can model efficiently geometrical or material discontinuities. When assessing crack initiation conditions in the bi-material notches, the generalized stress intensity factors H have to be calculated. Contrary to the determination of the K-factor for a crack in an isotropic homogeneous medium, for the ascertainment of the H-factor there is no procedure incorporated in the calculation systems. The calculation of these fracture parameters requires experience. Direct methods of estimation of H-factors need choosing usually length parameter entering into calculation. On the other hand the method combining the application of the reciprocal theorem (Ψ-integral) and FEM does not require entering any length parameter and is capable to extract the near-tip information directly from the far-field deformation.

Keywords: generalized stress intensity factor, fracture mechanics, bi-material notch, general singular stress concentrator

1. Introduction

The study of bi-material notches becomes a topical problem as they can model efficiently geometrical or material discontinuities. When assessing conditions for a crack initiation in the bi-material notches, the generalized stress intensity factors H are necessary to be calculated. In contrast to the determination of the K factor for a crack in an isotropic homogeneous medium, for the ascertainment of a generalized stress intensity factor (GSIF) there is no procedure incorporated in the calculation systems. The calculation of these fracture mechanics parameters is not trivial and requires certain experience. Nevertheless the accuracy of the H-factors calculation directly influences the reliability of assessment of the stress concentrators. Direct methods of estimation of H factors require choosing usually length parameter entering into calculation. On the other hand the method combining the application of the reciprocal theorem (Ψ-integral) and FEM does not require entering any length parameter and is capable to extract the near-tip information directly from the far-field deformation where the numerical fields are more accurate. The latter method can be readily applied to bi-materials composed of orthotropic materials components. In the paper various methods of calculation of the GSIFs are presented, tested and mutually compared. Recommendations for reliable evaluation of critical conditions of the bi-material notches are suggested.

*Corresponding author. e-mail: klusak@ipm.cz.

2. Bi-material notch

The model of a bi-material notch is suitable to simulate a number of construction points from which a failure is initiated. In the case of layered or fibre composite locations where the layers or fibres touch the surface of the composite body the singular stress concentrations occur. The singular stress distribution is derived on the basis of Airy stress functions in the form of Williams' expansion [10]. In most of the geometrical and material configurations of a bi-material notch there are two terms of the expansion with the real stress singularity exponents p_1 and p_2 in the interval $(0; 1)$. Contrary to a crack in a homogeneous material, the exponents differ from 1/2 and, furthermore, each singular term includes both normal and shear mode of loading, see [4] for detail. Then the singular stress components can be written in the polar coordinates:

$$\sigma_{mij} = \frac{H_1}{\sqrt{2\pi}} r^{-p_1} F_{ij1m} + \frac{H_2}{\sqrt{2\pi}} r^{-p_2} F_{ij2m}, \tag{1}$$

where for $\{i, j\} = \{r, \theta\}$ and for each singular term $k = 1, 2$:

$$
\begin{aligned}
F_{rrkm} &= (2 - p_k)(-a_{mk}\sin((2 - p_k)\theta) - b_{mk}\cos((2 - p_k)\theta) + \\
&\quad + 3c_{mk}\sin(-p_k\theta) + 3d_{mk}\cos(-p_k\theta)) \\
F_{\theta\theta km} &= (p_k^2 - 3p_k + 2)(a_{mk}\sin((2 - p_k)\theta) + b_{mk}\cos((2 - p_k)\theta) + \\
&\quad + c_{mk}\sin(-p_k\theta) + d_{mk}\cos(-p_k\theta)) \\
F_{r\theta km} &= (2 - p_k)(-a_{mk}\cos((2 - p_k)\theta) + b_{mk}\sin((2 - p_k)\theta) + \\
&\quad + c_{mk}\cos(-p_k\theta) - d_{mk}\sin(-p_k\theta))
\end{aligned}
$$

The subscript m differentiates the materials 1 and 2 where the stresses are determined. The values H_k are the generalized stress intensity factors that follow from the numerical solution of the studied geometry with given materials and boundary conditions [4, 3, 7]. The numerical calculation of the values H_k is necessary step for the final determination of the stress distribution. Numerical approaches to calculation of GSIFs have varying level of difficulty and accuracy. In the following paragraphs the direct methods and the method combining the application of the reciprocal theorem and FEM are described.

3. Direct methods of the generalized stress intensity factors H_k determination

Direct methods compare the results of some appropriate magnitude from a numerical solution with its analytical representation.

3.1. Tangential stress

The tangential stress $\sigma_{\theta\theta}$ is used here as the appropriate magnitude for the comparison. If the stress distribution is described by a combination of H_1 and H_2, it is necessary to solve the system of two equations. To achieve this, the values of $\sigma_{\theta\theta}$ following from the finite element method are determined for two different directions θ_1, θ_2. Then knowing the analytical relations e.g. for $\sigma_{\theta\theta}$ (1) we solve the system of equations for H_1 and H_2:

$$
\begin{bmatrix}
r^{-p_1} F_{\theta\theta 1m}(\theta = \theta_1) & r^{-p_2} F_{\theta\theta 2m}(\theta = \theta_1) \\
r^{-p_1} F_{\theta\theta 1m}(\theta = \theta_2) & r^{-p_2} F_{\theta\theta 2m}(\theta = \theta_2)
\end{bmatrix}
\begin{bmatrix}
H_1 \\
H_2
\end{bmatrix}
=
\begin{bmatrix}
\sigma_{m\theta\theta}(r, \theta_1) \\
\sigma_{m\theta\theta}(r, \theta_2)
\end{bmatrix}, \tag{2}
$$

The valid values of H_1, H_2 are then determined by an extrapolation of the solutions (2) into $r = 0$. The dependence of GSIF H_k on the polar coordinate r and the extrapolation into the notch vertex is shown in fig. 1.

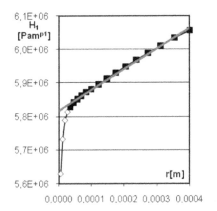
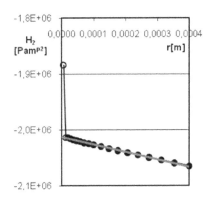

Fig. 1. Extrapolation of H_1 and H_2 values into the notch tip ($r = 0$)

3.2. Generalized strain energy density factor

For combined mode of loading it is suitable to use strain energy density (SED) approach to describe crack behaviour and to estimate the GSIFs as well. In the early 70s Sih [9] showed that damaging of a material could be estimated using strain energy density factor S that is defined by following equation:

$$S = r \cdot dW/dV = r \int_0^\varepsilon \sigma \, d\varepsilon, \tag{3}$$

In the same way the generalized strain energy density factor (SEDF) Σ is defined for the case of a bi-material notch. Limiting only to plane problems we can obtain the relation for the distribution of the SEDF Σ that in contrast to a crack, depends on the radial distance r. Because of the dependency of Σ on r, mean value over a certain distance d will be considered in the following [4]. For material m it is:

$$
\begin{aligned}
\overline{\Sigma_m} &= \frac{1}{d} \int_0^d \Sigma_m dr \\
&= \frac{H_1^2}{16 G_m \pi} \left(\frac{d^{1-2p_1}}{2 - 2p_1} U_{1m} + \frac{d^{1-2p_2}}{2 - 2p_2} h_{21}^2 U_{2m} + \frac{d^{1-p_1-p_2}}{2 - p_1 - p_2} 2 h_{21} U_{12m} \right),
\end{aligned}
\tag{4}
$$

where

$U_1 = [(F_{rr1m}^2 + F_{\theta\theta1m}^2)(k_m + 1) + 4F_{r\theta1m}^2 + 2F_{\theta\theta1m}F_{rr1m}(k_m - 1)]$

$U_2 = [(F_{rr2m}^2 + F_{\theta\theta2m}^2)(k_m + 1) + 4F_{r\theta2m}^2 + 2F_{\theta\theta2m}F_{rr2m}(k_m - 1)]$

$U_{12} = [(F_{rr1m}F_{rr2m} + F_{\theta\theta1m}F_{\theta\theta2m})(k_m + 1) + 4F_{r\theta1m}F_{r\theta2m} +$
$\qquad + (F_{\theta\theta1m}F_{rr2m} + F_{\theta\theta2m}F_{rr1m})(k_m - 1)]$

and $k_m = (1 - \nu_m)/(1 + \nu_m)$ for plane stress and $k_m = (1 - 2\nu_m)$ for plane strain, G_m is shear modulus and ν_m is the Poisson ratio. $h_{21} = H_2/H_1$ denotes ratio of the generalized stress intensity factors.

The integration distance d enters the calculations as a structural parameter or a parameter related to the mechanism of rupture.

The values H_1 and H_2 can be determined from the mean value of the SEDF. Here the two unknown parameters are calculated from the two following conditions. First the ratio h_{21} is

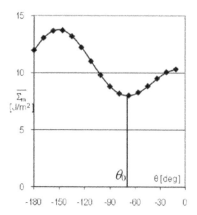

Fig. 2. Numerically gained distribution of the mean value of the generalized strain energy density factor with its minimum in the direction θ_0

gained from the numerically ascertained angle θ_0 of minimum of the mean value of the generalized SEDF $\overline{\Sigma_m}$, see fig. 2. For the minimum of $\overline{\Sigma_m}$ it must be satisfied the condition of the first derivation equal to zero:

$$\left(\frac{d^{2-2p_1}}{2-2p_1} \frac{\partial U_{1m}}{\partial \theta} + \frac{d^{2-2p_2}}{2-2p_2} h_{21}^2 \frac{\partial U_{2m}}{\partial \theta} + \frac{d^{2-p_1-p_2}}{2-p_1-p_2} 2h_{21} \frac{\partial U_{12m}}{\partial \theta} \right)_{\theta_0} = 0 , \qquad (5)$$

From two possible solutions h_{21} to the quadratic equation (5) we take only one that satisfies positive the second derivation, and thus it implies the minimum.

Finally knowing the ratio $h_{21} = H_2/H_1$ the value of GSIF H_1 is determined as:

$$H_1 = \left[\frac{1}{\overline{\Sigma_m}8dG_m} \left(\frac{d^{2-2p_1}}{2-2p_1} U_{1m} + \frac{d^{2-2p_2}}{2-2p_2} h_{21}^2 U_{2m} + \frac{d^{2-p_1-p_2}}{2-p_1-p_2} 2h_{21} U_{12m} \right) \right]^{\frac{-1}{2}} . \qquad (6)$$

The value $\overline{\Sigma_m}$ in (6) results from the numerical solution; d is a distance at which the value of SEDF is approximated. The value of d has to be chosen considering the failure mechanism. For the brittle fracture d relates to a grain size and thus it can express the increment of a crack. On the other hand for the fatigue loading d can be chosen as the plastic zone size. The value of the coefficient H_2 is determined reciprocally from: $H_2 = h_{21}H_1$.

4. Ψ-integral method

GSIF can also be determined using the so-called Ψ-integral [2]. This method is an implication of the Betti's reciprocity theorem which in the absence of the body forces states that the following integral is path independent:

$$\Psi(\mathbf{u}, \mathbf{v}) = \int_\Gamma [\sigma_{kl}(\mathbf{u})n_k v_l - \sigma_{kl}(\mathbf{v})n_k u_l] \, ds , \quad k, l = 1, 2. \qquad (7)$$

The contour Γ surrounds the notch tip and $\mathbf{u}, \mathbf{v}$ are two admissible displacement fields. Major advantage of the integral (7) is its path independency for the case of the multimaterial wedges.

To apply the Ψ-integral, it is convenient to derive the displacement field $\mathbf{u}$ and $\mathbf{v}$ using the Lekhnickii-Eshelby-Stroh (L.E.S.) formalism which allows expressing displacements and resultant forces $\mathbf{T}$ along the material interfaces from the complex potential theory. For the case of the orthotropic bi-material notch, one can write

$$u_{m,i} = 2\text{Re}\left\{\sum_{j=1}^{2} A_{m,ij}f_{m,j}(z_{m,j})\right\}, \quad T_{m,i} = -2\text{Re}\left\{\sum_{j=1}^{2} B_{m,ij}f_{m,j}(z_{m,j})\right\}, \quad (8)$$

where

$$\mathbf{A}_m = \begin{bmatrix} s_{m,11}\mu_{m,1}^2 + s_{m,12} & s_{m,11}\mu_{m,2}^2 + s_{m,12} \\ s_{m,12}\mu_{m,1} + s_{m,22}/\mu_{m,1} & s_{m,12}\mu_{m,2} + s_{m,22}/\mu_{m,2} \end{bmatrix},$$

$$\mathbf{B}_m = \begin{bmatrix} -\mu_{m,1} & -\mu_{m,2} \\ 1 & 1 \end{bmatrix}, \quad (9)$$

$s_{m,ij}$ are elements of the compliance matrix, $\mu_{m,i}$ are the material eigenvalues and m differentiates the materials of the notch. The potentials $f_{m,j}(z_{m,j})$ are considered in the form

$$f_{m,j}(z_{m,j}) = \phi_{m,j}z_{m,j}^{1-p}, \quad j, m = 1, 2, \quad (10)$$

where p is the stress singularity exponent and $z_{m,j} = x + \mu_{m,j}y$. The compatibility equations are automatically satisfied by (8) and the application of the boundary conditions of the joint, i.e. the stress free conditions along the notch faces and the displacement and stress continuity conditions along the bimaterial interface, leads to the eigenvalue problem whose solution are eigenvectors $\phi_{m,i}$ corresponding to the eigenvalue (exponent) p. It can be proved [8], that if $p, \phi_{m,i}$ is the solution of the eigenvalue problem mentioned above, it exists the so-called auxiliary solution $p^*, \phi_{m,j}^*$ of the same eigenvalue problem, where $p^* = 2 - p$. If the displacements $\mathbf{u}$ and $\mathbf{v}$ in (7) are chosen so that e.g. $\mathbf{u}$ corresponds to the regular solution $p, \phi_{m,i}$ and $\mathbf{v}$ to the auxiliary solution $p^*, \phi_{m,j}^*$ and vice versa, the Ψ-integral is nonzero. The other combination of the solutions $\mathbf{u}$ and $\mathbf{v}$ gives zero value of the Ψ-integral.

The isotropic materials are from L.E.S. point of view degenerated because the complex numbers $\mu_{m,1} = \mu_{m,2} = i$ are double roots of the characteristic equation of each material m. The complex coordinates $z_{m\cdot i} = x + \mu_{m,i}y$ reduce to the single value $z = x + iy$ and the matrices $\mathbf{A}_m$ and $\mathbf{B}_m$ are singular so that L.E.S. representation is unable to define the various fields. In this circumstance, it is useful to introduce Mushkhelishvili's complex potentials $\varphi(z)$ and $\psi(z)$ which allow to express the displacement field $\mathbf{u}$ and resultant forces $\mathbf{T}$ as follows

$$\begin{aligned} -2iG_m(u_{m,1} + iu_{m,2}) &= \kappa_m\varphi_m(z) - (z - \bar{z})\overline{\varphi_m'(z)} - \overline{\psi_m(z)} \\ T_{m,1} + iT_{m,2} &= \varphi_m(z) - (z - \bar{z})\overline{\varphi_m'(z)} - \overline{\psi_m(z)}, \end{aligned} \quad (11)$$

where $\kappa_m = 3 - 4\nu_m$ for plane strain and $(3 - \nu_m)/(1 + \nu_m)$ for plane stress, ν_m and G_m are Poisson's ratio and shear modulus of material m, respectively. With a view to relate the potentials $\varphi(z)$ and $\psi(z)$ in (11) with $f_{m,j}(z_{m,j})$ in (8), the equation (11) can be rewritten into the form (8), [1], where

$$\mathbf{A}_m = \frac{1}{4G_m i}\begin{bmatrix} \kappa_m i & -i \\ \kappa_m & 1 \end{bmatrix}, \quad \mathbf{B}_m = \frac{1}{2}\begin{bmatrix} i & -i \\ 1 & 1 \end{bmatrix}, \quad (12)$$

and

$$\begin{aligned} f_{m,1}(z) &= \varphi(z) = \phi_{m,1}z^{1-p}, \\ f_{m,2}(z) &= \psi(z) + (\bar{z} + z)\varphi'(z) \\ &= \phi_{m,2}z^{1-p} + (\bar{z} + z)(1 - p)\phi_{m,1}z^{-p} \quad m = 1, 2. \end{aligned} \quad (13)$$

Following (7) the displacement field anywhere around the notch tip can be written as

$$\mathbf{u}_m(r,\theta) = \mathbf{u}_m(0) + H_1 r^{1-p_1}\mathbf{u}_{m,1}(\theta) + H_2 r^{1-p_2}\mathbf{u}_{m,2}(\theta) + \dots, \tag{14}$$

where $\mathbf{u}_m(0)$ is the rigid body motion, dots represent non-singular terms of the expansion, r and θ are polar coordinates and $r^{1-p_1}\mathbf{u}_{m,1}(\theta)$ and $r^{1-p_2}\mathbf{u}_{m,2}(\theta)$ are basis functions corresponding to the coefficients H_1 and H_2, respectively, derived from (10) for the case of orthotropic materials or from (13) for the case of isotropic materials. Due to the orthogonality conditions that satisfy regular and auxiliary solutions the GSIFs H_1 and H_2 can be computed as follows

$$H_1 = \frac{\Psi(\mathbf{u}_m, r^{2-p_1^*}\mathbf{u}_1^*)}{\Psi(r^{2-p_1}\mathbf{u}_1, r^{2-p_1^*}\mathbf{u}_1^*)}, \quad H_2 = \frac{\Psi(\mathbf{u}_m, r^{2-p_2^*}\mathbf{u}_2^*)}{\Psi(r^{2-p_2}\mathbf{u}_2, r^{2-p_2^*}\mathbf{u}_2^*)}. \tag{15}$$

Since the exact solution $\mathbf{u}_m(r,\theta)$ in (14) is not known, a finite element solution can be used as an approximation for $\mathbf{u}_m(r,\theta)$ to obtain an approximation for H_1 and H_2.

5. Stability criteria suggestion

For the final assessment of a construction with a bi-material notch it is necessary to check the stability of the notch. Determination of the stability conditions of notches means to find the external loading under which a crack is initiated in the notch tip. The classic fracture mechanics approach of comparison of the stress intensity factor K_I with its critical value K_{Icrit} (represented by fracture toughness K_{IC} or by the fatigue threshold value K_{Ith}) is generalized to the following relation:

$$H_k(\sigma_{\text{appl}}) < H_{k\text{crit}}(M_m). \tag{16}$$

The value $H_k(\sigma_{\text{appl}})$ follows from the numerical solution and its determination is described in the previous paragraphs. The critical value $H_{k\text{crit}}$ depends on the critical material characteristic K_{IC} or K_{Ith} and has to be deduced with help of a controlling variable L, see [6]. The detail of derivation of critical values of the H-factor can be found in [4].

Then the critical applied stress is gained from the critical value of $H_{k\text{crit}}$:

$$\sigma_{\text{crit}} = \sigma_{\text{appl}}\frac{H_{1\text{crit}}}{H_1(\sigma_{\text{appl}})}. \tag{17}$$

Where σ_{appl} is the external loading stress applied in the numerical solution for the value H_1. The crack will not be initiated in the bi-material wedge tip if the applied stress is lower than the critical stress:

$$\sigma_{\text{appl}} < \sigma_{\text{crit}}. \tag{18}$$

6. Numerical example

The numerical study is performed on the rectangular bi-material notch loaded as shown in figure 3.

Within the numerical study the methods of calculation of GSIF were tested for varying combination of the material components expressed by Young's moduli. The results of both presented direct methods are compared. Fig. 4 shows the dependence of the values of GSIFs H_1 and H_2 on the ratio of moduli $E_1/E_2 \in \langle 0.012\,5; 10\rangle$.

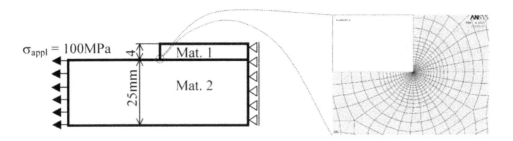

Fig. 3. Rectangular bi-material wedge used in the numerical example, a detail of a FEM mesh

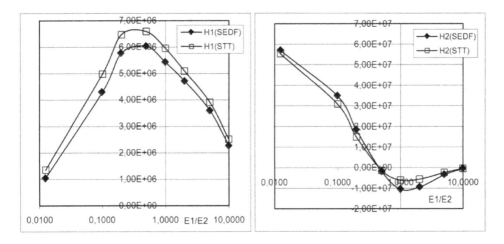

Fig. 4. Values of generalized stress intensity factors H_1, H_2 gained from the mean value of the generalized strain energy density factor (SEDF) and from the tangential stress (STT)

7. Conclusions

The results of GSIFs ascertained by means of two presented direct methods on the basis of two variables match well each other. Both methods give reliable input parameters to assessment of composite structures. The advantage of the direct methods is that they can be easily used with standard finite element calculation systems. The method gaining GSIFs in the two steps – from the supposed angle of potential crack initiation and from the mean value of the particular variable – can be easily programmed and thus automated. The two-step approach described here by means of mean value of generalized SEDF can be analogically derived for the mean value of tangential stress as well.

On the other hand the direct methods require choice of usually length parameter entering into calculation. It can be the choice of the region of the linear part suitable for the extrapolation of H-factor values (section 3.1) or the choice of the region for averaging the SEDF values. This handicap can be exploited for the optimization of the numerical processes by allowing entering structural parameters (e.g. grain size) or parameters connected with the loading type (e.g. plastic zone size, a crack increment) into calculations.

The method combining the application of the reciprocal theorem (Ψ-integral) and FEM is capable to extract the near-tip information directly from the far-field deformation where the numerical fields are more accurate. Thus the Ψ-integral method can help with the option of the unknown length parameter. Further the latter method can be readily applied to bi-materials composed of orthotropic materials components.

Note that for the final evaluation of the bi-material notch it is further necessary to determine critical applied stress on the basis of critical values of the generalized stress intensity factors. These approaches are described within suggestion of stability criteria, see e.g. [4]. When the critical applied loading is determined it is suitable to keep the same controlling variable for the GSIFs estimation as well as for the stability criterion suggestion.

Acknowledgements

The authors are grateful for financial support through the Research project of Academy of Sciences of the Czech Republic AVOZ20410507 and the project of Czech Science Foundation 101/08/0994.

References

[1] Choi, S. T., Shin, H., Earmme, Y. Y., On the unified approach to anisotropic and isotropic elasticity for singularity, interface an crack in dissimilar media, International Journal of Solids and Structures, Vol. 40, (2003), 1 411–1 431.

[2] Desmorat, R., Leckie, F. A., Singularities in bi-materials: parametric study of an isotropic/anisotropic joint. European Journal of Mechanics – A/Solids, Vol. 17, (1998), 33–52.

[3] Hilton, P. D., Sih, G. S., In Mechanics of Fracture – Methods of Analysis and Solutions of Crack Problems, by G. C. Sih, Noordhoff International Publishing, Leyden, p. 426–477, (1973).

[4] Klusák, J., Knésl, Z., Náhlík, L., Crack initiation criteria for singular stress concentrations, Part II: Stability of sharp and bi-material notches, Engineering mechanics 14 (6) (2007) ISSN 1802-1484, 409–422.

[5] Knésl, Z., A Criterion of V-notch Stability, International Journal of Fracture 48 (1991) R79–R83.

[6] Knésl, Z., Klusák, J., Náhlík, L., Crack Initiation Criteria for Singular Stress Concentrations, Part I: A Universal Assessment of Singular Stress Concentrations, Engineering Mechanics 14 (6) (2007) ISSN 1802-1484, 399–408.

[7] Owen, D. R. J., Fawkes, A. J., Engineering Fracture Mechanics: Numerical Methods and Applications. Pineridge Press Ltd, Swansea, (1983).

[8] Papadakis, P. J., Babuska, I., A numerical procedure for the determination of certain quantities related to the stress intensity factors in two-dimensional elasticity. Computer Methods in Applied Mechanics and Engineering Vol. 122, (1995), pp. 69–92.

[9] Sih, G. C., A special theory of crack propagation, In G. C. Sih (Ed.), Mechanics of Fracture – Methods of analysis and solutions of crack problems, Noordhoff International Publishing, Leyden, The Netherlands, 1973, XXI–XLV.

[10] Williams, M. L. The stresses around a fault or crack in dissimilar media, Bull. Seismol. Soc. Amer., 49 (2) (1959) 199–204.

Reconstruction of a fracture process zone during tensile failure of quasi-brittle materials

V. Veselý[a,*], P. Frantík[a]

[a] *Faculty of Civil Engineering, BUT Brno, Veveří 331/95, 602 00 Brno, Czech Republic*

Abstract

The paper outlines a technique for estimation of the size and shape of an inelastic zone evolving around a crack tip during the tensile failure of structures/structural members made of quasi-brittle building materials, particularly cementitious composites. The technique is based on an amalgamation of several concepts dealing with the failure of structural materials, i.e. multi-parameter linear elastic fracture mechanics, classical non-linear fracture models for concrete (equivalent elastic crack and cohesive crack models), and the plasticity approach. The benefit of this technique is expected to be seen in the field of the determination of fracture characteristics describing the tensile failure of quasi-brittle silicate-based composites. The method is demonstrated using an example of a (numerically simulated) fracture test involving the three-point bending of a notched beam and validated by experimentally obtained results of three-point bending and wedge-splitting tests taken from the literature.

Keywords: cementitious composites, fracture, quasi-brittle, fracture process zone, multi-parameter linear elastic fracture mechanics, effective crack model, cohesive crack model, Rankine strength criterion

1. Introduction

Fracture of quasi-brittle materials is accompanied by formation and evolution of an inelastic zone around the propagating crack tip. This zone, referred to as the fracture process zone (FPZ), is responsible for the non-linear manner of the quasi-brittle fracture. In this zone the material is damaged via various failure mechanisms on several levels of material structure. The FPZ size can not be neglected in relation to the size of the structures/structural members made of quasi-brittle materials; it may be comparable in size to the material's basic constituents (e.g. the aggregate size in the case of cementitious composites) or be greater in extent by up to two orders of magnitude [5, 11, 21, 24]. The existence of this zone influences the values of fracture-mechanical properties determined by some of the standardized evaluation procedures from records of fracture tests, e.g. those based on the work-of-fracture method [18], as was extensively reported in literature (for the most relevant see [2, 7, 9, 12, 23]). This is caused by the fact, according to the opinion of the authors, see e.g. [29, 19], that the characteristics of the FPZ do not enter these evaluating procedures.

For determination of either the propagating crack tip position or rather the entire FPZ extension during the fracture process several experimental techniques have been reported in the literature (summarized e.g. in [21, 24]). Holographic interferometry technique was used in combination with digital image analysis to determine the FPZ extent from specimen surface strain fields [21, 24]. Other reported successful techniques utilize such phenomena as acoustic emission [14, 17, 15], X-ray radiation [24, 17], and infrared thermography [24]. Recently, a method based on micro/nano-indentation of side surfaces of a cracked specimen has been tried out [30] in order to disclose the spatial (surface) distribution of mechanical properties of the tested material from which the extent of the material damage zone accumulated during the fracture process can be estimated.

In the paper an analytical method is shown which enables modelling of the FPZ during the fracture process in quasi-brittle materials, and the estimation of its size and shape. The energy dissipated within the FPZ during fracture can then be related to the extent of this zone of failure, which can result in the refining and better specification of the procedures of determination of fracture-mechanical characteristics of quasi-brittle materials, especially fracture energy [18]. In this paper, application of the method is illustrated in the context of fractures of notched beams under three-point bending and wedge-splitting test specimens.

2. Theoretical background

2.1. Classical non-linear models for quasi-brittle fracture

As was already mentioned above, the fundamental characteristic of the tensile failure of quasi-brittle materials is the existence of the FPZ around the crack tip. This phenomenon is the reason for the non-linear fracture behaviour of these materials. Non-linear models developed for quasi-brittle fracture are divided into two basic groups according to the approach to its capture.

The simplest class of these models, which are known as equivalent elastic crack models (EECM), simulate the cohesive fracture of quasi-brittle materials by replacing the real body, which contains a crack of a certain initial length and a FPZ ahead of it, with a brittle body with an effective (sharp) crack longer than the initial one, and then forcing both bodies to exhibit the same structural behaviour. The essential advantage of these models is the linear elastic fracture mechanics (LEFM) apparatus preserved for analyses within these models. The effective crack model [16] is the representative of this group utilized in the presented technique. The effective crack length in the loaded body is determined based on the change in the secant compliance of the body between the initial and the current stage of the fracture process.

A more advanced class of models which consider the mutual cohesive effect of crack faces in the vicinity of the crack tip resulting in non-linear fracture behaviour are referred to as cohesive crack models (CCM). According to this approach a crack and the FPZ evolving around the crack tip in a quasi-brittle body is modelled by an extension Δa of the original crack of length a. The crack faces are clamped by cohesive forces on a section of the cohesive crack extension from the crack tip up to a point where the crack opening displacement is critical. The progress of fracture modelled by the cohesive crack approach can be described as follows: If the external loading of the body causes such a stress distribution at the original crack tip that the stress component opening the crack exceeds the tensile strength of the material, the original crack starts to propagate and the newly formed faces of the crack extension Δa are compressed by cohesive stress $\sigma(w)$, where w is the crack opening. By this process a realistic situation is modelled in which an FPZ arises and starts to develop at the initial crack tip. When the current crack opening w at the initial crack tip position exceeds a critical value w_c, the whole cohesive zone starts to shift forward through the body. This situation corresponds to the separation of the FPZ from the original crack tip and the beginning of its movement through the body. An increment of the stress-free crack remains behind the FPZ. The technique for the reconstruction of FPZ size and shape described hereinafter employs tools used in the fictitious crack model by Hillerborg et al. [8], which is a representative of cohesive crack models for quasi-brittle materials. Nowadays, the main applications of these models are their implementations into FEM codes developed for the design and assessment of concrete and reinforced concrete structures (e.g. ATENA [6], DIANA, etc.).

2.2. *Multi-parameter fracture mechanics*

Williams' solution for an elastic isotropic homogeneous 2D body with a crack [32, 13] provides an approximation of stress and deformation fields within the body by means of its expansion into a power series. For a stress tensor it holds:

$$\sigma_{ij} = \sum_{n=1}^{\infty} \left(A_n \frac{n}{2} \right) r^{\frac{n}{2}-1} f_{ij}(n,\theta) , \tag{1}$$

where r and θ are the polar coordinates, coefficients A_n are known constants and f_{ij} are known functions of the angle θ. A closer look at the three stress components reveals the following form:

$$\begin{Bmatrix} \sigma_x \\ \sigma_y \\ \tau_{xy} \end{Bmatrix} = \sum_{n=1}^{\infty} \left(A_n \frac{n}{2} \right) r^{\frac{n}{2}-1} .$$

$$\begin{Bmatrix} \left[2 + (-1)^n + \frac{n}{2} \right] \cos \left(\frac{n}{2} - 1 \right) \theta - \left(\frac{n}{2} - 1 \right) \cos \left(\frac{n}{2} - 3 \right) \theta \\ \left[2 - (-1)^n - \frac{n}{2} \right] \cos \left(\frac{n}{2} - 1 \right) \theta + \left(\frac{n}{2} - 1 \right) \cos \left(\frac{n}{2} - 3 \right) \theta \\ -\left[(-1)^n + \frac{n}{2} \right] \sin \left(\frac{n}{2} - 1 \right) \theta + \left(\frac{n}{2} - 1 \right) \sin \left(\frac{n}{2} - 3 \right) \theta \end{Bmatrix} . \tag{2}$$

The first term of the series (1) or (2), respectively, is singular with regard to the distance r from the crack tip (it tends to infinity for $r \to 0$) and the constant A_1 from this term corresponds to the stress intensity factor K. The second term is constant with respect to the value of r and is referred to as T-stress. The rest of the terms take finite values for arbitrary r.

For an approximate description of the stress and deformation fields in the very vicinity of the crack tip (the asymptotic fields) it is possible to consider only the first term (which is the case of classical fracture mechanics) or the first two terms (the case of two-parameter fracture mechanics) of the series and ignore the terms for $n > 2$, as they converge to zero for $r \to 0$. This is possible because the failure of an ideally brittle material starts at a single point — the crack tip — for which the asymptotic description holds true.

However, in the case of a quasi-brittle material the FPZ arises and evolves around the crack tip; the size of this zone can not be neglected in relation to the dimensions of the cracked body (including the crack length). And if one needs to describe the more distant surroundings of the crack tip, higher order terms of the Williams' series must be taken into account. The FPZ size substantially exceeds a range in which the stress state can be described accurately enough only by means of classical fracture mechanics or even two-parameter fracture mechanics. Therefore, the approach when more terms of the Williams' series are used to describe stress and displacement fields in the cracked body can be referred to as an application of multi-parameter LEFM.

3. Estimation of the plastic zone in metallic materials

The size and shape of the plastic zone in elasto-plastic materials influences crack behaviour substantially. For an estimation of its extent two-parameter fracture mechanics (the description of the near-crack-tip fields performed by means of K and T — elastic — or J and Q — elasto-plastic [13]) is usually utilized, as its size is relatively small (small scale yielding). In the procedure of plastic zone contour calculation an equivalent (comparative) measure of the stress field near the crack tip, determined e.g. from principal stresses with the use of the Mises or Tresca yield criterion, is compared with a particular critical value [1, 20], which is the value of yield stress or yield stress in shear, respectively. Eq. (2) can be rearranged into a closed form that for a particular angle θ explicitly returns the radius r. An example of the extent of plastic zones for both plane stress and plane strain conditions is depicted in fig. 1.

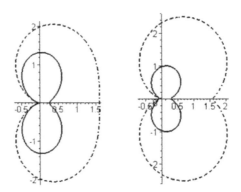

Fig. 1. Plastic zone estimation for the plane stress (dashed line) and plane strain (solid line) states for Mises (left) and Tresca (right) yield criteria (courtesy of S. Seitl [20])

4. Estimation of the fracture process zone in quasi-brittle cementitious materials

For the estimation of the fracture process zone in quasi-brittle materials a concept similar to that outlined in the previous section is proposed. However, essential modifications are neces-sary. The first concerns the size of the (plastic) zone. Cementitious materials are characterized by much lower tensile strength than metallic materials, and this results in a plastic zone[1] of much greater extent. The second reason is the cohesive manner of the quasi-brittle fracture. Cohesive stress is transferred between crack faces with an adequately small opening displace-ment. Therefore, the fracture process zone is much larger also in the direction opposite to the crack propagation, and this is due to the cohesive forces caused by various fracture mechanisms (taking place at the (nano-,) micro- and mezo-level).

A detailed description of the proposed method for FPZ extent estimation is performed in [27, 28]. Here we only highlight the fundamental ideas/approaches on which the technique is primarily based:

- *Equivalent elastic crack models* [16, 11] are used for the estimation of the effective crack tip during the fracture process. The effective crack model enables calculation of the effec-tive crack length in the cracked body from the change in compliance of the body between the beginning and the current stage of the fracture process. The iterative procedure de-scribed in detail in [25] is employed.

---

[1]The term 'plastic' is used here because of the similarity of the concept with that described in sec. 3. In reality, the failure mechanisms in quasi-brittle cementitious composites are of a different nature (e.g. microcracking, aggregate interlock, crack bridging and branching etc. [21, 11, 24]).

- *Multi-parameter linear elastic fracture mechanics* [32, 13] is used for a description of the stress field in the specimen (not necessarily only in the very vicinity of the crack tip). Within the more distant surroundings of the crack tip, the consideration of which is essential for quasi-brittle materials with a large FPZ, higher order terms of Williams' series (eq. (2)) are taken into account.

- *The theory of plasticity* [10, 31] is employed for the determination of the zone of material where the equivalent stress exceeds the tensile strength. The Rankine failure criterion is used in this paper; however, other criteria suitable for cementitious composites, e.g. those of Drucker-Prager or Willam-Warnke [31] are possible. The contour of the Rankine equivalent stress profile is determined using the Newton's iteration method in this procedure. Direct solution of the radius r and angle θ might not be appropriate if more than two terms of Williams' series are used (for up to two terms of the series the radius r can be explicitly expressed as a function of the angle θ).

- *Cohesive crack models* [8, 21, 5] are taken into account when introducing the cohesive law into the procedure of FPZ range estimation. This approach aids in capturing the non-linear behaviour of (cohesive) material in the FPZ. The FPZ is assembled as a union of plastic zones determined at points on the face of the propagating crack (i.e. previous crack tip positions) where the crack opening displacement is lower than its critical value.

These approaches are used within the processing of fracture test records, typically load-displacement diagrams (P–d diagrams, see e.g. fig. 3 right), in the following sequence:

i) For individual stages of the fracture process, i.e. the points of the P–d diagram, the length of the equivalent elastic crack is estimated by means of the effective crack model.

ii) The stress state in a body with an effective crack is approximated through Williams' power series; the number of terms in the series must be chosen with respect to the mutual relation between the assumed FPZ size/shape and the size/shape of the body (with respect to the distance between the FPZ and the free boundaries of the body).

iii) The extent of the zone where the until-now elastic material starts to fail, denoted here as Ω_{PZ} (the initials 'PZ' in the subscript stands for Plastic Zone), is determined by comparing the tensile strength f_t of the material to a proper characteristics of the stress state around the crack tip (some sort of equivalent stress σ_{eq}, for cementitious composites e.g. the Rankine failure criterion can be employed). As was already mentioned above, the estimation of the plastic zone boundary is performed iteratively using the Newton's method.

iv) The crack opening displacement profile for each crack tip position is calculated via appropriate LEFM formulas, e.g. from [22].

v) In agreement with the cohesive crack approach, the FPZ is supposed to extend from the zone of the current failure around the current crack tip (at stage i, denoted as $\Omega_{PZ,i}$), where the selected stress state characteristic (equivalent stress σ_{eq}) exceeds the tensile strength f_t, up to a point on the crack faces where the value of crack opening displacement reaches its critical value (i.e. the value of the cohesive stress drops to zero). This point corresponds to a prior stage of the fracture process (let's denote it by the subscript k, and the zone of failure corresponding to that stage as $\Omega_{PZ,k}$). For a given stage i the FPZ is considered to be a union of zones Ω_{PZ} with indices from the interval $\langle i, k \rangle$.

In this paper two alternative FPZ expressions are considered. In the first case, the FPZ is denoted as $\Omega_{\text{FPZ,basic}}$; the union of the individual plastic zones is performed simply without any other treatment. In the second case, $\Omega_{\text{FPZ,scaled}}$, the zones for the individual indices (corresponding to points on the crack face, i.e. locations of the effective crack tip during prior stages of fracture) are scaled by a factor corresponding to the relative value of the cohesive stress for those points, i.e. by $\sigma(w)/f_{\text{t}}$. The envelope of Ω_{PZ} zones for all stages of the fracture represents a region in which some sort of damage has taken place during the fracture process throughout the entire specimen ligament. It is denoted here as Ω_{WRAP}. This method of FPZ definition for the current crack tip is based on the assumption that the energy dissipation in the failure processes occurs at those points in the body where the equivalent stress σ_{eq} appropriate to prior stages of the fracture has exceeded the tensile strength f_{t} (failure mechanisms have started to proceed there).

The construction of an FPZ evolving during fracture in quasi-brittle materials is illustrated in fig. 2. The thick solid lines in the graph indicate the initial crack face, and the front and back boundaries of the specimen. The other lines correspond to the boundaries of the zones described above: the plastic zone $\Omega_{\text{PZ},i}$ for the current crack tip position (point i); the plastic zone $\Omega_{\text{PZ},k}$ for the last crack tip position (point k), where the cohesive stress is still active at the i-th stage of the fracture process; the fracture process zone (i.e. the union of plastic zones within the range of action of the cohesive stress) in two possible variants of its representation (basic vs. scaled); and the wrap (i.e. the union of plastic zones for all crack tip positions along the specimen ligament — the area where the material undergoes damage).

The procedure is being developed as a JAVA application under the name ReFraPro (Reconstruction of Fracture Process) [26].

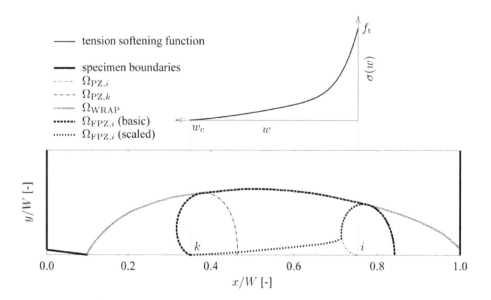

Fig. 2. FPZ construction: typical cohesive law for quasi-brittle materials (top); boundaries of the above-described formulations of zones of failure (bottom)

5. Examples

5.1. Numerically simulated test of a Single-Edge Notched beam under Three-Point Bending (SEN-TPB)

Let us consider a notched beam of dimensions $L \times W \times B$ equal to $0.48 \times 0.08 \times 0.08$ m with an initial crack/notch of length $a_0 = 0.008$ m (relative notch length $\alpha_0 = a/W = 0.1$) loaded in three-point bending (see fig. 3 left). The beam is supposed to be made of concrete with the following fracture-mechanical parameters: tensile strength $f_t = 3.7$ MPa, fracture energy $G_F = 93$ Jm^{-2} and exhibiting exponential (Hordijk's, see e.g. [6]) softening.

The fracture test of the beam was simulated numerically using ATENA software [6] for this illustrative example. The recorded load-displacement diagram is plotted in fig. 3 right, where six stages of the fracture process are emphasized (A to F) which correspond to relative effective crack lengths α equal to 0.15, 0.35, 0.53, 0.7, 0.87, and 0.97 (respectively).

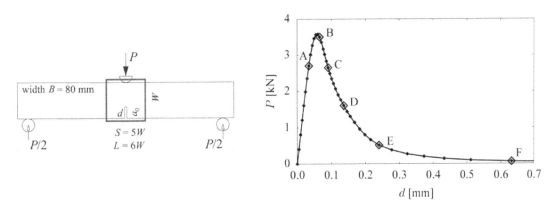

Fig. 3. SEN-TPB testing configuration (left), numerically simulated P–d diagram for the above-described beam including values for six stages of the fracture process (right)

In fig. 4, the central part of the SEN-TPB specimen (highlighted in fig. 3 left) is displayed with a depiction of the evolution of the following quantities/phenomena corresponding to stages of the fracture process indicated in the P–d diagram, fig. 3 right:

- Equivalent stress distribution over the specimen (left column). From this picture one can estimate the extent of the plastic zone, i.e. the zone of material damage that initiates at the current load step. The stress level corresponding to the value of tensile strength, which creates the boundary of the plastic zone, is displayed in black.

- Extent of the fracture process zone, i.e. of the area where the cohesive stress is active. Both representations of the zone considered in this paper (i.e. basic and scaled) are depicted (middle columns). The intensity of the cohesive stress is indicated by colour (gray) scale — the lightest colour indicates that the cohesive stress value is equal to tensile strength; the darkest corresponds to zero cohesive stress.

- The extent of the area where the material undergoes damage, denoted here as 'WRAP' (right column). It is created as an envelope of the plastic zones in all stages of the fracture.

For the construction of the zones four terms of the power series (see equation (2)) (the values of coefficients A_1 to A_4 for their evaluation were taken from [13]) and the Rankine

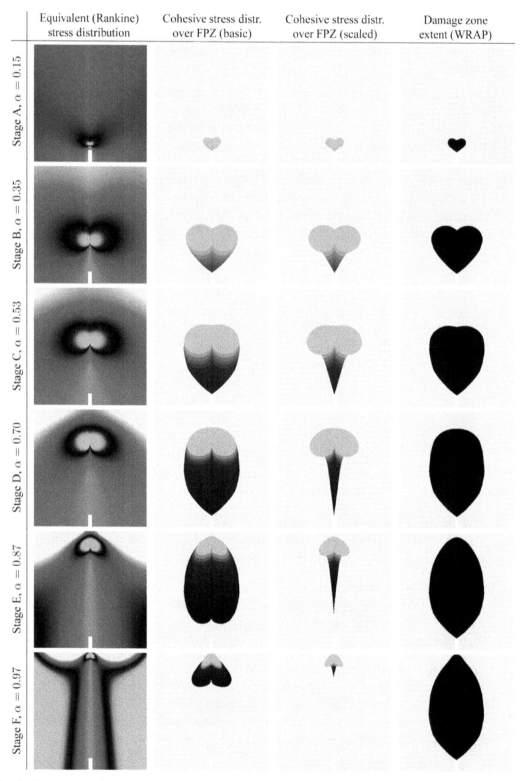

Fig. 4. Central part of the SEN-TPB specimen displaying the evolution of the equivalent (Rankine) stress distribution, the FPZ extent in both the basic and the scaled expression with indications of the intensity of the cohesive stress over the FPZ, and the WRAP extent, all for the six stages of the fracture process indicated in the load-displacement diagram in fig. 3

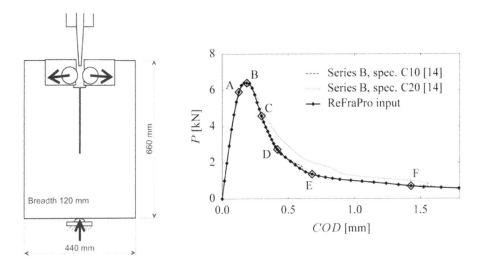

Fig. 5. Experimental set-up and dimensions of WST [14] (left) and load-displacement diagram (right)

failure criterion (in a plane stress state) are taken into account — such settings were proved to be applicable in previous studies by the authors [27, 28].

5.2. *Experiments on Wedge-Splitting Test (WST) and SEN-TPB specimens*

Two sets of tests were selected for comparison with the results of the presented (semi-) analytical method from the rather modest amount of published works dealing with the experimental estimation of the zone of tensile failure in quasi-brittle cementitious composites. Techniques based on acoustic emission (AE) scanning were considered under the assumption that AE phenomena are similar to the approach employed within the proposed method.

The first set is the WST series conducted by Mihashi and Nomura [14]. They tested two sets of concrete and mortar specimens differing in strength and aggregate size. The experimental set-up with an indication of the specimens' dimensions is depicted in fig. 5 left. No P–d diagrams were reported in the paper and therefore we conducted our own numerical simulations in order to estimate the structural behaviour of the specimens, which serves as an input for the developed procedure. The numerical simulation was performed in ATENA software [6]; the material model was tuned according to the compressive strength of the material reported in the paper. Two specimens were selected for which the damage zone was reported in the paper: specimens marked as C10 and C20 (both from series B, maximum aggregate size equal to 10 and 20 mm, respectively) with compressive strength $f_c = 34.8$ and 30.5 MPa, respectively. The simulated P–d diagrams do not differ much, which is apparent from fig. 5, therefore a smoothed variant of the C10 P–d curve was used as the input for the procedure. The values of the other necessary parameters were taken as being the same as for the numerical simulation: tensile strength $f_t = 2.8$ MPa, fracture energy $G_f = 60$ Jm^{-2}, and Hordijk's exponential softening curve. The other settings were analogous to those in the previous example, e.g. four terms of Williams' series (taken from [13] for CT geometry) and the Rankine failure criterion in a plane stress state.

The progress of the FPZ extent with an indication of the intensity of the cohesive stresses within it[2] for six selected stages of the fracture process (emphasized in fig. 5 right) is shown in

---

[2]Here another scale is applied: From light grey corresponding to $\sigma_{coh} = f_t$, through dark gray, back to light gray corresponding to $\sigma_{coh} = 0$).

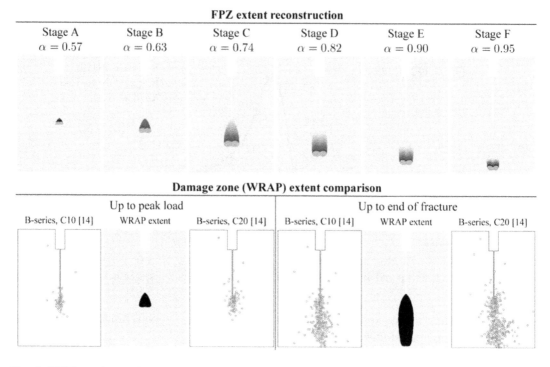

Fig. 6. WST specimen displaying the evolution of the FPZ extent (basic expression) for selected stages of the fracture process with indications of the cohesive stress intensity (top row); the WRAP extent for two stages of the fracture process in comparison to the damage zone estimated experimentally using AE [14]

the top row of fig. 6. The extent of the zones in which the AE sources were located (indicated by the small circles) in the two considered specimens by Mihashi and Nomura [14] is then compared to the WRAP extent in the bottom row of the fig. 6; for stage B (peak load) on the left, and for the end of fracture on the right. Very good agreement can be seen, especially for the final stage of the tests.

The second experimental data set was published by Muralidhara et. al [15]. They conducted SEN-TPB tests on two sizes of specimens accompanied again by AE scanning. From their series of experiments only one test (the specimen marked D2T20UB02) was reported in detail. The test configuration with the dimensions of the selected specimen is shown in fig. 7 top and the P–$CMOD$ curve (where $CMOD$ is the crack mouth opening displacement) present in their paper is plotted in fig. 7 left. From the reported P–d curve (where d is the mid-span deflection), see fig. 7 right, it is obvious that some problems have occurred during measurement of the deflection[3]. Therefore, numerical simulation in the ATENA software was again performed. The parameters of the material model were tuned so that the simulated and experimentally recorded P–$CMOD$ curves match. Using the relationship between these curves the 'true' experimental P–d curve was subsequently reconstructed, the smoothed version of which was then used as the input for the procedure for the estimation of the damage zone extent. The values of other

---

[3]The displacement was probably recorded using an LVDT sensor which was fitted to the frame of the testing machine instead of using a measuring frame fitted directly to the specimen. If this is the case, the part of the displacement occuring due to the pushing of the supports into the specimen is also recorded. However, only the part of the displacement corresponding to the crack propagation is relevant for the analysis.

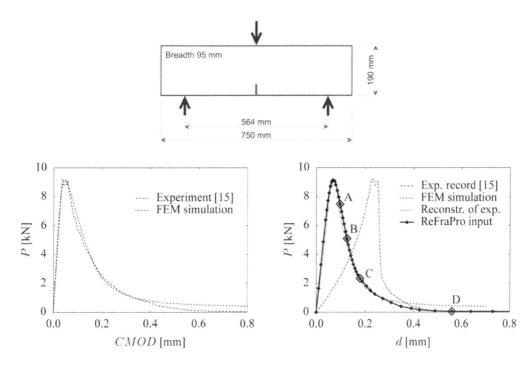

Fig. 7. Test configuration and dimensions of the SEN-TPB specimen [15] (top), load-crack mouth opening displacement diagram and load-deflection diagram, respectively (bottom)

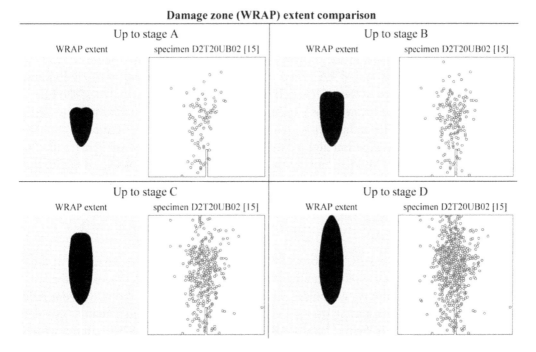

Fig. 8. Central part of the SEN-TPB specimen displaying the evolution of the WRAP extent for four selected stages of the fracture process in comparison to the experimentally estimated damage zone [15]

necessary inputs to the procedures were again equal to the values of the parameters of the material model used for the simulation: compressive strength $f_c = 42.5$ MPa, tensile strength $f_t = 4.3$ MPa, fracture energy $G_f = 90$ Jm^{-2}, and Hordijk's exponential softening curve. Four terms of Williams' series and the Rankine strength criterion were employed.

Fig. 8 shows a comparison between the predicted and the measured extent of the damaged zone at four stages of the test. The scatter of the AE event locations is higher whilst the density of the AE events is lower (especially for the initial stages of fracture) in the case of the Muralidhara et. al test in comparison to the Mihashi and Nomura tests. The experimentally estimated extent of the zone of cumulative damage was again predicted quite well.

6. Discussion of results

The first example serves as an illustration of some of the abilities of the developed procedure. The setting of the material parameter values (as the inputs to the procedure) in connection with rather small specimen dimensions results in estimation of considerable FPZ extent and WRAP width. In this case a relatively very accurate description of the stress state for the points far from the crack tip is necessary — it is obtained by using of at least 4 terms of Williams' series. In cases of larger specimens and/or more brittle materials 3 or even 2 terms may be sufficient. In contrast, for cases where the size of the FPZ takes even larger portion of the specimen's volume it is necessary to describe the stress field even more accurate.

Validation of the developed technique is performed only partially here, as the experimental data available in the literature (in most cases AE data) usually provide the zones of cumulative failure through the entire fracture process. Such zones correspond to the envelopes of the FPZs referred to as WRAPs in this paper. As can be seen in figs. 6 and 8, the agreement in this aspect is reasonable.

However, a deeper analysis of the experimental data must be made, in particular that concerning the energetic demands of the individual AE events (which can influence the extent of the damage zone, both FPZ and WRAP) and concerning the assignment of the individual AE events to the individual steps of the test (which should reveal the current extent of the FPZ). For future work it is intended to locate some experiments on such aspects in the literature and/or to conduct them ourselves. The authors also plan to compare the outputs of the procedure to data obtained by other types of experimental technique, e.g. methods based on X-ray radiation or (nano-) micro-indentation. Verifications by means of numerical tools based on lattice models are currently being investigated by the authors as a complement to the experimental evidence.

7. Conclusions

In the paper a method is shown which provides an estimation of the size and shape of the fracture process zone, which is a typical feature accompanying the fracture process in quasi-brittle materials. This method employs a combination of various approaches from different fields of theory of fracture mechanics and plasticity, yields reasonable results and contains considerable potential. This technique is being developed in order to create/refine procedures which enable the determination of fracture parameters of quasi-brittle materials independent of the size, shape and boundary conditions of laboratory test specimens. The procedure shall relate the energy dissipated in the FPZ to its volume and is currently under intensive investigation by the present authors.

Acknowledgements

This outcome has been achieved with the financial support of the Ministry of Education, Youth and Sports, project No. 1M0579, within the activities of the CIDEAS research centre. In this undertaking, theoretical results gained in the project of the Czech Science Foundation, project No. GA 103/07/P403, were partially exploited.

References

[1] Anderson, T. L., Fracture mechanics: Fundamentals and Applications, Third Edition, CRC Press, 2004.

[2] Bažant, Z. P., Analysis of work-of-fracture method for measuring fracture energy of concrete, J. Engrg. Mech. (ASCE), 122(2) (1996) 138–144.

[3] Bažant, Z. P., Kazemi, M. T., Determination of fracture energy, process zone length and brittleness number from size effect, with application to rock and concrete, Int. J. Fract. 44 (1990) 111–131.

[4] Bažant, Z. P., Oh, B. H., Crack band theory for fracture of concrete, Mater. Struct. 16 (1983) 155–177.

[5] Bažant, Z. P., Planas, J., Fracture and size effect in concrete and other quasi-brittle materials, CRC Press, Boca Raton, 1998.

[6] Červenka, V. et al., ATENA Program Documentation, Theory and User Manual, Cervenka Consulting, Prague, 2005.

[7] Duan, K., Hu, X.-Z., Wittmann, F. H., Boundary effect on concrete fracture and non-constant fracture energy distribution, Engrg. Fract. Mech. 70 (2003) 2 257–2 268.

[8] Hillerborg, A., Modéer, M., Petersson, P.-E., Analysis of crack formation and crack growth in concrete by means of fracture mechanics and finite elements, Cem. Concr. Res. 6 (1976) 773–782.

[9] Hu, X.-Z., Duan, K., Influence of fracture process zone height on fracture energy of concrete, Cem. Concr. Res. 34 (2004) 1 321–1 330.

[10] Jirásek, M., Zeman, J., Deformation and failure of materials — viscoelasticity, plasticity, fracture and damage (in Czech), Czech Technical University in Prague, 2008.

[11] Karihaloo, B. L., Fracture mechanics and structural concrete, Longman Scientific & Technical, New York, 1995.

[12] Karihaloo, B. L., Abdalla, H. M., Imjai, T., A simple method for determining the true specific fracture energy of concrete, Mag. Concr. Res. 55 (2003) 471–481.

[13] Knésl, Z., Bednář, K., Two-parameter fracture mechanics: Calculation of parameters and their values (in Czech), Institute of Physics of Materials, Czech Academy of Sciences, 1998, Brno.

[14] Mihashi, H., Nomura, N., Correlation between characteristics of fracture process zone and tension-softening properties of concrete, Nuclear Engineering and Design 165 (1996) 359–376.

[15] Muralidhara, S., Raghu Prasad, B. K., Eskadri, H., Karihaloo, B. L., Fracture process zone size and true fracture energy of concrete using acoustic emission, Construction and Building Materials 24 (2010) 479–486.

[16] Nallathambi, P., Karihaloo, B. L., Determination of specimen-size independent fracture toughness of plain concrete, Mag. Concr. Res. 38 (1986) 67–76.

[17] Otsuka, K., Date, H., Fracture process zone in concrete tension specimen. Engrg. Fract. Mech. 65 (2000) 111–131.

[18] RILEM Committee FMT 50, Determination of the fracture energy of mortar and concrete by means of three-point bend test on notched beams, Mater. Struct. 18 (1985) 285–290.

[19] Řoutil, L., Veselý, V., Keršner, Z., Seitl, S., Knésl, Z., Fracture process zone size and energy dissipated during crack propagation in quasi-brittle materials, Proceedings of 17th European Congress on Fracture – ECF 2008 (book of abstracts + CD-ROM), J. Pokluda, P. Lukáš, P. Šandera, I. Dlouhý (eds.), Vutium, Brno, 2008, 97 + CD 8 p.

[20] Seitl, S., A study of the plastic zone within the framework of two-parameter fracture mechanics (in Czech), Proceedings of Computational Mechanics 2002, Nečtiny, 2002, 423–430.

[21] Shah, S. P., Swartz, S. E., Ouyang, C., Fracture mechanics of structural concrete: applications of fracture mechanics to concrete, rock, and other quasi-brittle materials, John Wiley & Sons, Inc., New York, 1995.

[22] Tada, H., Paris, P. C., Irwin, G. R., The stress analysis of cracks handbook, 3rd ed., Professional Engineering Publishing, Ltd., Bury St. Edmunds, UK, 2000.

[23] Trunk, B., Wittmann, F. H., Influence of size on fracture energy of concrete, Mater. Struct., 34 (2001) 260–265.

[24] van Mier, J. G. M., Fracture Processes of Concrete: Assessment of Material Parameters for Fracture Models, CRC Press, Inc., Boca Raton, 2007.

[25] Veselý, V., Parameters of concrete for description of fracture behaviour. PhD Thesis, Brno University of Technology, Faculty of Civil Engineering, Brno, 2004 (in Czech).

[26] Veselý, V., Frantík, P., *ReFraPro — Reconstruction of Fracture Process*, Java application, 2008.

[27] Veselý, V., Frantík, P., Development of fracture process zone in quasi-brittle bodies during failure, Proceedings of Engineering Mechanics 2009, J. Náprstek a C. Fischer (eds.), Svratka, Czech Rep., Institute of Theoretical and Applied Mechanics, v.v.i., Academy of Sciences of the Czech Republic, 288–289 + 1 393–1 404 (CD – in Czech).

[28] Veselý, V., Frantík, P., Keršner, Z., Cracked volume specified work of fracture, Proceedings of the 12th Int. Conf. on Civil, Structural and Environmental Engineering Computing, B. H. V. Topping, L. F. Costa Neves and R. C. Barros (eds.), Funchal, Civil-Comp Press, 2009.

[29] Veselý, V., Řoutil, L., Keršner, Z., Structural geometry, fracture process zone and fracture energy, Proceedings of Fracture Mechanics of Concrete and Concrete Structures (Proc. FraMCoS-6), Al. Carpinteri, P. Gambarova, G. Ferro, G. Plizzari (eds.), Catania, Italy, Taylor & Francis/Balkema, vol. 1 (2007) 111–118.

[30] Veselý, V., Keršner, Z., Němeček, J., Frantík, P., Řoutil, L., Kucharczyková, B., Estimation of fracture process zone extent in cementitious composites, Chem. Listy 104 (2010) 382–385.

[31] Wikipedia, the free encyclopedia — Stress (mechanics), Yield surface, http://en.wikipedia.org.

[32] Williams, M. L., On the stress distribution at the base of a stationary crack, ASME J. Appl. Mech. 24 (1957) 109–114.

11

Utilization of random process spectral properties for the calculation of fatigue life under combined loading

J. Svoboda[a,*], M. Balda[a], V. Fröhlich[a]

[a]Institute of Thermomechanics, Czech Academy of Sciences, Veleslavínova 11, 301 14 Plzeň, Czech Republic

Abstract

The contribution includes the results of experimental works aiming to find a new methodology for the calculation of fatigue life of structures subjected to operational loading from a combination of forces and moments of random character. Considering the fracture mechanics theory, then the damaging of material is both in the micro- and macro-plastic area connected with the rise of plastic deformation and hence with the plastic transformation rate which depends on the amount of supplied energy. The power spectral density (PSD) indicating the power at individual frequencies in the monitored frequency band yields information about the supplied amount of energy. Therefore, it can be assumed that there is a dependence between the PSD shape and the size of damage and that the supplied power which is proportional to the value of dispersion s^2 under the PSD curve could be a new criterion for the calculation of fatigue life under combined loading. The searching for links between the spectral properties of the loading process and the fatigue life of structure under load is dealt with by new Grant GA No. 101/09/0904 of the Czech Technical University in Prague and the Institute of Thermomechanics of the Academy of Sciences of the Czech Republic, v.v.i.

Keywords: power spectral density, random process, combined loading, plastic deformation, fatigue life

1. Introduction

The calculation of fatigue life of structures subjected to operational loading from a combination of forces and moments of random character is an issue the resolution of which is endeavored by prestigious sites in the area of fatigue worldwide for a number of years. Nevertheless, most of the structures in traffic and power engineering are subjected to exactly this general process of loading. Owing to the application of the "rain flow" method, which allows transforming a random loading process to a set of harmonic cycles and applications of Manson-Coffin relation, we are able nowadays to perform the calculation of fatigue life for the case of uniaxial loading. The hysteresis energy absorbed by material in one cycle, represented by an area of the closed hysteresis loop, has become a suitable criterion for the calculation of the size of damage. How to proceed, however, in the case of multiaxial random loading when in the decomposition of this operational stress by the above mentioned "rain flow" method the closed hysteresis loops do not develop at all? This way of loading will require finding another criterion how to project the continual changes of energy in the calculation of fatigue life.

The contribution will include the first results which have been obtained experimentally on tube specimens of material 11523.1 with a cross hole as a notch. The specimens were subjected to a combination of random tension-pressure and torsion loading with the PSD of a decreasing

---

*Corresponding author. e-mail: svoboda@cdm.it.cas.cz.

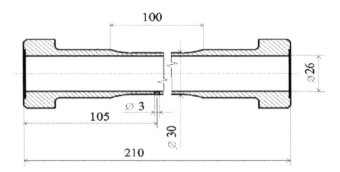

Fig. 1. Test specimen dimensions

shape and a shape of a pyramid. It appears that for the combination of random loading it is also possible to construct so called S-N curves as a dependence of standard deviations s_d of the resulting processes and the number N_b of loading blocks. All the fatigue life points lie on this curve regardless of the shape of partial processes. The more aggressive of the two monitored PSD was the pyramidal shape which is confirmed by the results from the monitoring of fatigue crack propagation.

2. Influence of PSD shape on fatigue life

The basic problem of the experimental research dedicated to the influence of the power spectral density on fatigue life is the ability to create a pseudorandom process with required statistical characteristics, probability density of its samples and power spectral density. For that purpose, it was necessary before starting the testing to develop several programs in the MATLAB environment which would allow to generate processes with the required PSD shapes (with predefined shapes of increasing, decreasing, constant and pyramidal courses) on one hand, and to calculate the setting parameters of machine INOVA for a couple of loading processes of force F and moment M_k by means of which the machine will be controlled in the planar loading regime on the other hand. The preparatory works are more in detail described in [1], and [2]. In order to increase the accuracy of results to be obtained, the software IFRM developed by firm INOVA was used to pin down the shape of applied random processes; the software assured that after several iterations the deviations of actual values in peaks were lower than 1 % from the requested ones. At the same time, the actual courses of force and moment, which were then used for the processing of information about the applied random processes statistical characteristics, were filed during the tests.

3. Experimental works and their results

The experimental tests aimed at the solution of the above issue were carried out using tube specimens 30 mm in diameter with a wall thickness of 2 mm and a notch shaped as a cross single-sided hole with a diameter of 3 mm – see fig. 1. The specimens were made of material ČSN 11523.1.

An electro-hydraulic testing machine ZUZ 200-1 made by firm INOVA was used for the testing allowing subjecting the specimens to a combination of tension-pressure and torsion loading. In addition to usual harmonic oscillation, the device also allows to apply blocks of random processes the time series of which are entered via a control computer of the machine in the form of special binary files.

Testing 10 specimens under a combined axial tension-pressure and torsion loading started the tests. Blocks of broadband random processes with a normal distribution in the frequency range 0–10 Hz were used for both the stress types. Both the independent processes differed in the PSD shape. A decreasing PSD shape was chosen for tension – pressure, and a shape of a pyramid for torsion.

The magnitude of both the processes was chosen so that the damaging effect of the two components, i.e. of normal and shear stresses, is equal, whereas the damaging effective stress defined as

$$\sigma_d(t) = \sigma(t) + ik_c\tau(t), \tag{1}$$

where $k_c = \sigma_c/\tau_c$ and i is the imaginary unit. Then, for the damaging stress dispersion

$$s_d^2 = s_\sigma^2 + k_c^2 s_\tau^2 \tag{2}$$

must apply that $s_\sigma^2 = k_c^2 s_\tau^2$. It follows from the equation (2) that the damaging stress standard deviation will be

$$s_d = s_\sigma\sqrt{2}. \tag{3}$$

The chosen number of samples in a sequence was 300 000, which corresponded to the duration T_1 of one block realization of 5 minutes.

Table 1. The effect of PSD shape on life

Spec. No.	Tension-pressure decreasing PSD. Torsion pyramidal PSD		Life N_b	Spec. No.	Tension-pressure pyramidal PSD Torsion decreasing PSD		Life N_b
	s_d [MPa]	s_d^2			s_d [MPa]	s_d^2	
1	125,36	15 715,6	48,40	11	130,636	17 065,66	45,64
2	122,81	15 082,7	55,43				
3	115,78	13 404,4	76,15	12	117,458	13 796,47	70,25
4	107,30	11 514,4	110,29				
5	102,22	10 449,6	159,72				
6	96,16	9 246,9	180,11	13	95,085	9 041,18	134,12
7	89,89	8 080,6	266				
8	83,05	6 897,8	434,96	14	84,240	7 096,38	376,96
9	76,59	5 865,6	765,16				
10	67,65	4 576,3	1 667,04	15	71,099	5 055,13	1 206,6

Tab. 1 shows the lives in the number of loading blocks N_b for 10 applied levels of combined stress related to the standard deviations of damaging stress of peaks s_d of the resulting loading process and a respective value of dispersion s_d^2.

In order to find how significantly the life of specimens is influenced by the PSD shape, we extended the tests by 5 more specimens the PSD shapes of which were exchanged for the forces and moments while the standard deviations of both the loading processes (their magnitudes) remained the same. Therefore, corresponding to the specimen No. 1 on the left side in the table is the specimen No. 11 on the right side, etc. It follows from tab. 1 that the exchange of PSD shape influenced the resulting life N_b. The shape of a pyramid in tension-pressure increased its aggression that is more evident from fatigue cracks monitoring.

In a similar way as in the cases of harmonic loading it is also possible to plot the S-N curves for the combined random loading. If we plot from the above table the dependence of the number of loading blocks N_b on the standard deviations s_d of damaging stress peaks of resulting processes, we will obtain the exponential dependence indicated in fig. 2.

As evident in the figure, the points marked with empty circles corresponding to the first ten specimens are close to the regression curve with only a very small dispersion although each of the points is a resultant of a different shape of randomly generated partial processes for the force F and moment M_k. It therefore means that it is not the order of loading levels of the partial processes which matters but the magnitude characterized by the standard deviations s_σ and s_τ which are used for the calculation of the process standard deviation or of the damaging stress peaks s_d. It is interesting that the points marked with full circles corresponding to the exchanged power spectral densities are also relatively close to the above regression curve. Only one from the above points deflects significantly from this curve with its error moving around 40 %.

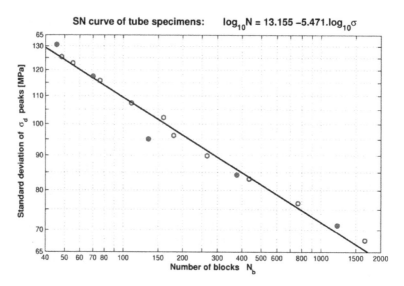

Fig. 2. Regresion curve of the tested specimens dependence $s_d - N_b$

4. Crack propagation under combined random loading

The crack propagation was monitored in specimens No. 3, 4, 5 and 6 subjected to a combination of random tension-pressure loading with a decreasing shape of PSD, and of torsion with a pyramidal PSD shape. The resulting process standard deviations s_d for these specimens are indicated in tab. 1.

In order to find how the exchange of the PSD shape of the applied random loadings (a pyramidal PSD shape in case of tension-pressure, and a decreasing shape in case of torsion) influences the propagation, we extended the works to specimens No. 12, 13 and 14. The propagation was monitored visually on the tube specimen surfaces. The results are indicated in fig. 3, 4, 5 and 6. It is evident from the figures that the cracks at the tube cross-hole were propagating at a given combination of random tension-pressure and torsion under all loading levels practically equally under an angle of about 30°. The differences were only in the propagation rate corresponding to the loading intensity.

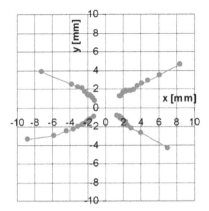

Fig. 3. Crack propagation in specimen No. 2

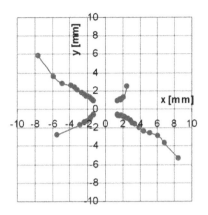

Fig. 4. Crack propagation in specimen No. 4

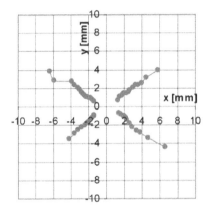

Fig. 5. Crack propagation in specimen No. 6

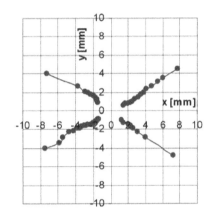

Fig. 6. Crack propagation in specimen No. 8

Fig. 7 shows the dependence of the above crack propagation rate on the range of stress intensity factor ΔK that was calculated from the formula

$$\Delta K = \Delta s_d \sqrt{\pi a_s}, \tag{4}$$

where Δs_d is the range of standard deviations of the resulting processes peaks and the crack length a_s corresponds to the mean value of 4 cracks shown in fig. 3–6. This dependence corresponds to the known Paris formula that can be entered in the form

$$da_s/dN_b = C \Delta K^n, \tag{5}$$

where N_b is the fatigue life in the number of loading blocks for the monitored specimens No. 2, 4, 6 and 8. The values of C and n constants of the above Paris formula (4) for the 4 monitored tube specimens are shown in tab. 2.

Fig. 7 shows the dependences da/dN_b on the ΔK_{ef} for the monitored 4 test specimens No. 2, 4, 6 and 8 and also the regression curve passing through all the plotted points. Its equation is indicated in the fig. 7.

For the case of our combined tension-pressure – torsion loading of the test specimens, it is not necessary to relate the stress intensity factor only to the standard deviation of the loading resulting process peaks but to the standard deviation of the loading partial process peaks. Here,

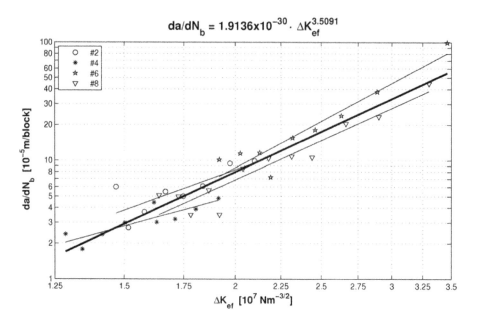

Fig. 7. Dependence of da_s/dN_b for the monitored 4 tube specimens

Table 2. Paris formula constants obtained by the application of formulas (6) and (7)

Specimen No.	Application of formula (5)		Application of formula (7)	
	C	n	C	n
2	2,601 7E–34	4,041 3	2,565 3E–34	4,044 4
4	5,274 4E–30	3,436 3	5,574 3E–30	3,436 3
6	9,118 2E–25	2,727 3	1,020 1E–24	2,727 3
8	4,613 6E–20	2,058 3	4,732 7E–20	2,058 3

we start from the assumption that during loading, the crack tip displacement is caused by a combination of at least 2 modes (in the given case mode I. and mode II.). In this case, however, it is necessary to operate with the effective value of the stress intensity factor for the calculation of which the ref. [3] shows several relations. According to Chen and Keer, it applies that

$$\Delta K_{ef} = \sqrt[8]{(\Delta K_I^2 + 3\Delta K_{II}^2)^3(\Delta K_I^2 + \Delta K_{II}^2)}, \tag{6}$$

where $\Delta K_I = \Delta s_\sigma \sqrt{\pi a_s}$ and $\Delta K_{II} = \Delta s_\tau \sqrt{\pi a_s}$.

The values Δs_σ and Δs_τ are the ranges of standard deviations of partial processes (tension-pressure and torsion) peaks. The Paris formula will then have a form of

$$\frac{da_s}{dN_b} = C\Delta K_{ef}^n. \tag{7}$$

The constants C and n are again indicated in tab. 2. If we compare both the approaches for the calculation of Paris formula, we can see that the constants C and n in both the cases do not differ too much. Therefore, the differences in the calculation of residual life for the area of crack propagation will be small.

Let us have a look now, what will be with the crack origin and propagation in the same tube specimen if it is subjected to a combination of random tension-pressure and torsion with

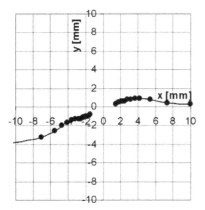

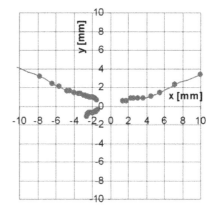

Fig. 8. Crack propagation in specimen No. 12   Fig. 9. Crack propagation in specimen No. 13

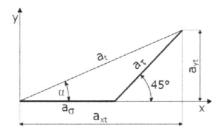

Fig. 10. Damage vectors under combined loading (theory)

the exchanged PSD shapes (tension-pressure – a pyramid, torsion – decreasing). It can be seen from the above tab.1 that the fatigue lives N_b of the specimens No. 11–15 were decreased which indicates that the resulting effect of the combined random loading acts more aggressively. What is interesting, it is the way of crack propagation at a given way of loading. While in case of the specimens No. 2, 4, 6 and 8 the cracks originated at the tube specimen cross hole were propagating in all 4 directions almost uniformly under an angle of about $30°$, with the exchange of PSD shapes the way of crack propagation significantly changed as evident in fig. 8 and 9. By increasing the tension-pressure aggressiveness, due to the assignment of a more aggressive PSD shape (a pyramid) and by decreasing the torsion aggressiveness by choosing a decreasing PSD shape, the damaging effect of torsion was suppressed and the propagation shape approximated the propagation under uniaxial tension-pressure. The original incline angle of $30°$ decreased to a value of around $20°$. The level of torsion stress (a value of the torsion process standard deviation s_τ) will obviously play its role here as well.

As evident from the above, the PSD shape is an important informer about the aggressiveness of the process of damaging a structure under dynamic stress and is significantly applied in connection with the character and level of the applied loading independently of the loading being uniaxial or combined. Hence, the different aggressiveness of the loading process can be characterized based on the PSD shapes, and the areas under these curves represent the size of the supplied power proportional to dispersion s^2.

Based on the above crack propagation monitoring we can have an idea of the damaging under a combined tension-pressure – torsion loading. Let us assume a crack of length a_t, see fig. 10, which originated at a combination of these loadings when the tension-pressure process had a decreasing PSD shape and the torsion a pyramidal shape. As experimentally discovered, the crack was propagating under an angle of $\alpha_e = 30°$. With respect to the requirement for the same damage from the loading processes we can assume that the vectors of damage from normal stress (tension-pressure) and shear stress will be of the same size according to fig. 10, they will, however, differ in the direction. The vector of damage from normal stress will cause damage in the x-axis direction, the shear stress vector under an angle of $45°$ as evident from the fig. 10.

If a_{xt} and a_{yt} are crack length components a_t, the following will apply

$$a_{xt} = a_\sigma + a_\tau \cdot \cos 45° \tag{8}$$

$$a_{yt} = a_\tau \cdot \sin 45°, \tag{9}$$

where a_σ and a_τ are the vectors of damage from normal and shear stresses.

The following applies for the angle α in the fig. 10

$$\operatorname{tg} \alpha = \frac{a_{yt}}{a_{xt}} = \frac{a_\tau \sin 45°}{a_\sigma + a_\tau \cos 45°} = \frac{\sin 45°}{\frac{a_\sigma}{a_\tau} + \cos 45°}. \tag{10}$$

With respect to the requirement for the same damage from normal and shear stresses, it must apply that $a_\sigma = a_\tau$, i.e. the equation (1) will have a form of

$$\operatorname{tg} \alpha = \frac{\sin 45°}{1 + \cos 45°} = 0{,}414\,2. \tag{11}$$

Corresponding to this value is the angle $\alpha = 22{,}5°$.

The experimentally found value of the propagating cracks incline in fig. 3–6 moves somewhat higher around the value of $\alpha = 30°$ which means that the actual component of damage from shear stress will be obviously slightly higher than that from the normal stress as evident in fig. 11.

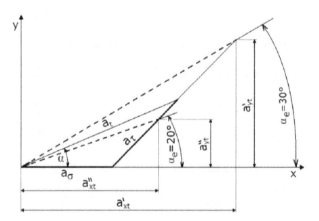

Fig. 11. Damage vectors under combined loading (experiments)

How will it be now, in the other case, when we exchanged the PSD shapes of both the loading processes? It can be seen in fig. 8 and 9 that the character of cracks arisen at the hole has changed. The angle α decreased to about $20°$ and the way of crack propagation shows evidence of suppressing the damaging effect of torsion and the increase of damage from the tension-pressure component that is also evident in fig. 11. Lessening the angle α will also change the ratio a_τ/a_σ which will be < 1. It can be presumed from the above that by the exchanging of the PSD shapes, the damaging effect of both the processes changed and that the different PSD shapes manifesting themselves by different aggression can under combined loading significantly influence not only the process of fatigue damage and hence the resulting fatigue life (see tab. 1) but also the character of originated cracks and the progress of their propagation.

5. Summary of partial results

The results of the performed experimental works monitoring the influence of the power spectral density shape on the fatigue life of structures under combined random tension-pressure and torsion loading can be briefly summarized into the following points:

- In a similar way as in the case of uniaxial loading, it is also possible to construct the S-N curves for the combined random loading if for the values S we put the standard deviations s_d of the resulting process or its peaks, and for the value N we put the fatigue life in the number of loading blocks N_b. Decisive for the fatigue life is a so called "process magnitude" expressed by the value s_d, not the course of random processes.

- The partial loading processes PSD shape influences the resulting fatigue life. Every shape manifests itself by different aggression. Nevertheless, this aggression shall always be connected with the mode of stress (the same PSD shape will influence the resulting life differently in case of tension-pressure, torsion or bending).

- The PSD shape of individual stress components also influences the direction of fatigue cracks propagation. In our case, the exchange of a less aggressive decreasing shape for a more aggressive pyramidal shape of tension-pressure caused the lessening of the propagating crack incline angle by about $10°$, from the original $30°$ to $20°$.

- The known Paris formula can be applied to determine the rate of fatigue crack propagation. The values of constants C and n depend on the intensity of loading, not on the method of calculation of the stress intensity factor range. Almost identical values of these constants were obtained for the case when for the calculation of ΔK we used a range of the standard deviation Δs_d of the stress resulting process peaks, or a procedure according to Chen and Keer, where for the calculation of value ΔK_{ef} we put the ranges of the peaks standard deviations Δs_σ and Δs_τ of individual stress components.

- The contribution of individual stress components to the total damage can be determined for the known direction of fatigue crack propagation by vector superposition. We can assume for our case of tension-pressure and torsion combination that the damaging caused by tension will be perpendicular to the direction of the tube specimen longitudinal axis, and by torsion under an angle of $45°$. The level of partial damage can be determined according to the crack propagation angle.

6. Conclusion

The contribution summarizes partial results of the experimental works the aim of which was to obtain information about the linkage between the PSD shape of applied combined random stress components (in this case tension-pressure and torsion) and the resulting fatigue life in the number of loading blocks N_b. Two PSD shapes, i.e. a combination of a decreasing shape and a pyramidal shape, were monitored in a frequency range of 0-10 Hz. The applied random processes differed from each other but had the same damaging affect. The pyramidal shape both of tension-pressure and torsion proved to be more aggressive. Further experimental works will be aimed at the combinations of a constant shape (white noise), an increasing shape, possibly some other. The obtained results will then be used for the formulation of energy criterion for the calculation of fatigue life.

Acknowledgements

The research has been supported by Grant No. 101/09/0904 of the Czech Science Foundation and by the ASCR project No. CEZ: AV0Z 20760514.

References

[1] Balda, M., Loading processes with various spectral properties for fatigue tests, Proc. of Conf. Engineering Mechanics 2006, Svratka, ITAM ASCR, eds. Náprstek, Fischer, Paper 20, ISBN 978-80-86246-35-2.

[2] Svoboda, J., Balda, M., Influence of spectral properties of combined random loading on fatigue life, Proc. of Conf. Engineering Mechanics 2006, Svratka, ITAM ASCR, eds. Náprstek, Fischer, Paper 20, ISBN 978-80-86246-35-2.

[3] Wei-Ren Chen, Keer, L. M., Journal of Engineering Materials and Technology, Vol. 113, 1991, pp. 222–227.

12

Numerical simulation of plastic deformation of aluminium workpiece induced by ECAP technology

R. Melicher[a]

[a] Faculty of Applied Mechanics, Žilina university in Žilina, Univerzitná 1, 010 26 Žilina, Slovak Republic

Abstract

The objective of this paper was to show some numerical simulations which can be very helpful for optimal settings in extrusion process by equal channel angular pressing (ECAP) technology. Using the finite element (FEM) software ADINA all basic types of used shape of the inner and outer corner of die were modeled. The influence of the tool geometry for plastic strain by simple shear into the aluminum workpiece during extrusion process of the ECAP was analyzed. It was also examined the influence of the pressing force for all usually used variants of die corners.

Keywords: equal channel angular pressing, finite element simulation, accumulative effective plastic strain, effective stress, pressing force

1. Introduction

In recent years, bulk nanostructured materials (NSM) processed by methods of severe plastic deformation (SPD) have attracted the growing interest of specialists in material science. This interest is conditioned not only by unique physical and mechanical properties inherent to various NSM, e.g. processed by gas condensation or high energy ball milling with subsequent consolidation but also by several advantages of SPD materials as compared to other NSM.

Segal and co-workers developed the method of ECA pressing realizing deformation of massive billets via pure shear in the beginning of 80s. Its goal was to introduce intense plastic strain into materials without changing the cross-section area of billets. Due to that, their repeat deformation is possible. In the early 90s the method was further developed and applied as an (SPD) method for processing of structures with submicron and nanometric grain sizes [1].

Methods of SPD should meet a number of requirements, which are to be taken into account while developing them for formation of nanostructures in bulk samples and billets. These requirements are as follows [1, 2]:

- Firstly, it is important to obtain ultra-fine grained structures with prevailing high angle grain boundaries since only in this case can a qualitative change in properties of materials.

- Secondly, the formation of nanostructures uniform within the whole volume of a sample is necessary for providing stable properties of the processed materials.

- Thirdly, though samples are exposed to large plastic deformations they should not have any mechanical damage or cracks.

*Corresponding author. e-mail: richard.melicher@fstroj.uniza.sk,

Traditional methods of SPD, such as rolling, drawing or extrusion cannot meet these requirements. Formation of nanostructures in bulk samples is impossible without application of special mechanical schemes of deformation providing large deformations at relatively low temperatures as well as without determination of optimal regimes of material processing. At present the majority of the obtained results are connected with application of two SPD methods: severe plastic torsion straining under high pressure (SPTS) and equal channel angular pressing (ECAP) [1].

It is well established that SPD is useful for grain refinement in metallic materials. There are so many SPD processes available for grain refinement: for instance, equal channel angular pressing (ECAP), high pressure torsion (HPT), multiple forging and accumulative roll bonding (ARB). A common feature to all these SPD processes is that the cross-sectional dimensions of samples remain unchanged after deformation so that it is possible to introduce large strain into the samples [1, 2].

Of these various procedures, ECAP is an especially attractive processing technique for several reasons:

- First, it can be applied to fairly large billets so that there is the potential for producing materials that may be used in a wide range of structural applications.

- Second, it is a relatively simple procedure that is easily performed on a wide range of alloys and, except only for the construction of the die, processing by ECAP uses equipment that is readily available in most laboratories.

- Third, ECAP may be developed and applied to materials with different crystal structures and to many materials ranging from precipitation-hardened alloys to intermetallics and metal-matrix composites.

- Fourth, reasonable homogeneity is attained through most of the as-pressed billet provided the pressings are continued to a sufficiently high strain.

- Fifth, the process may be scaled-up for the pressing of relatively large samples and there is a potential for developing ECAP for use in commercial metal-processing procedure.

These various attractive features have led to many experimental studies and new developments in ECAP processing over the last decade.

2. Fundamental parameters in ECAP

2.1. Slip systems for the different processing routes

During ECA pressing a billet is multiple pressed through a special die using an ECA facility in which the angle of intersection of two channels is usually 90°. If necessary, in the case of a hard-to-deform material, ECAP is conducted at elevated temperatures.

Since the cross-sectional area remains unchanged, the same sample may be pressed repetitively to attain exceptionally high strain. For example, the use of repetitive pressings provides an opportunity to invoke different slip systems on each consecutive pass by simply rotating the samples in different ways between the various passes [1, 2, 3, 4, 5].

In practice, many of the investigations of ECAP involve the use of bars with square cross-sections and dies having square channels. For these samples, it is convenient to develop processing routes in which the billets are rotated by increments of 90° between each separate pass. The same processing routes are also easily applied when the samples are in the form of rods with the circular cross-sections. During ECAP the direction and number of billet passes through the channels are very important for microstructure refinement. In papers and books the following routes of billets were considered (see fig. 1) [1, 2, 5]:

- route A where the sample is pressed repetitively without any rotation,
- route B_A where the sample is rotated by $90°$ in alternate directions between consecutive passes,
- route B_C where the sample is rotated in the same sense by $90°$ between each pass,
- route C where the sample is rotated by $180°$ between passes.

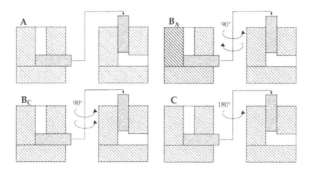

Fig. 1. The four fundamental processing routes in ECAP

The given routes are distinguished in their shear directions at repeat passes of a billet through intersecting channels. Due to that, during ECAP a change in a spherical cell within a billet body occurs.

The different slip systems associated with these various processing routes are depicted schematically in fig. 2 where X, Y and Z planes correspond to the three orthogonal planes and slip is shown for different passes in each processing route – thus, the planes labelled 1 through 4 correspond to the first 4 passes of ECAP.

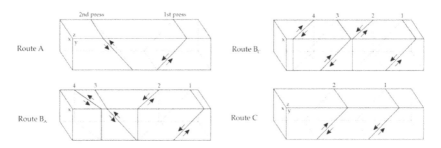

Fig. 2. The slip systems viewed on the X, Y and Z planes for consecutive passes using processing routes A, B_A, B_C and C

In route C, the shearing continue on the same plane in each consecutive passage through the die but the direction of shear is reversed on each pass – thus, route C is termed a redundant strain process and the strain is restored after every even number of passes. It is apparent that route B_C is also a redundant strain process because slip in the first pass is cancelled by slip in the third pass and slip in the second pass is cancelled by slip in the fourth pass. By contrast, routes A and B_A are not redundant strain processes and there are two separate shearing planes intersecting at an angle of $90°$ in route A and four distinct shearing planes intersecting at angles of $120°$ in route B_A. In routes A and B_A, there is a cumulative build-up of additional strain on each separate pass through the die.

2.2. Analytical solution of the effective plastic strain

Iwahashi et al. have proposed, in 1996, a relation between effective strain ε_{eff} and the ECAP angles ϕ (value of the angle within the die between the two parts of the channel) and ψ (value of the angle at the outer arc of curvature where the channels intersect). This expression is calculated from von Mises isotropic yield criterion applied to the pure shear condition [1, 2]

$$\varepsilon_{eff} = \frac{1}{\sqrt{3}} \left[2 \cot \left(\frac{\phi}{2} + \frac{\psi}{2} \right) + \psi \csc \left(\frac{\phi}{2} + \frac{\psi}{2} \right) \right]. \tag{1}$$

The effective strain ε_{eff} in component form is represented by

$$\varepsilon_{eff} = \left[\frac{2 \left[\varepsilon_x^2 + \varepsilon_y^2 + \varepsilon_z^2 + \left(\left(\gamma_{xy}^2 + \gamma_{yz}^2 + \gamma_{zx}^2 \right) / 2 \right) \right]}{3} \right]^{1/2}. \tag{2}$$

Since the same strain is accumulated in each passage through the die, the effective strain after N extrusion passes ε_N may be expressed in a general form by the relationship

$$\varepsilon_N = N \cdot \varepsilon_{eff}. \tag{3}$$

Thus, the strain may be estimated from this equation for any pressing condition provided the angles ϕ and ψ are known.

2.3. Analytical solution of the extrusion pressure

Alkorta et al. proposed an upper-bound solution to the ECAP pressure considering Hollomon-type materials and using a frictionless condition. According to these authors, the pressure is related with the material hardening behaviour as [1, 2, 5]

$$P = \left(\frac{\sigma_y}{n+1} \right) \left[\frac{2 \cot \left((\phi + \psi) / 2 \right) + \psi}{\sqrt{3}} \right]^{(n+1)} \tag{4}$$

where σ_y and n are yield stress and the strain-hardening exponent, respectively.

An analogue ECAP pressure solution considering a Swift-type material is based on

$$\bar{\sigma} = K \left(\varepsilon_0 + \bar{\varepsilon}^p \right)^n \tag{5}$$

where K is the strength coefficient, ε_0 is the pre-strain and $\bar{\varepsilon}^p$ is the effective von Mises plastic strain. Thus, considering one pass of extrusion the equation (4) become

$$P = \left(\frac{K}{n+1} \right) \left\{ \left(\frac{\sigma_y}{K} \right)^{\left(\frac{1}{n} \right)} + \left[\frac{2 \cot \left((\phi + \psi) / 2 \right) + \psi}{\sqrt{3}} \right]^{(n+1)} - \left[\left(\frac{\sigma_y}{K} \right)^{\left(\frac{1}{n} \right)} \right]^{(n+1)} \right\}. \tag{6}$$

The extrusion force per unit of thickness after one pass can be obtained multiplying the right side of the equation (6) by the width (W) of the billet. Thus

$$\frac{P}{thickness} = W \left(\frac{K}{n+1} \right) \left\{ \left(\frac{\sigma_y}{K} \right)^{\left(\frac{1}{n} \right)} + \left[\frac{2 \cot \left((\phi + \psi) / 2 \right) + \psi}{\sqrt{3}} \right]^{(n+1)} - \right.$$
$$\left. - \left[\left(\frac{\sigma_y}{K} \right)^{\left(\frac{1}{n} \right)} \right]^{(n+1))} \right\}. \tag{7}$$

3. Finite element simulation of ECAP process

Nowadays finite element method is used for simulation of technological processes increasingly. FEM simulations are helpful to estimate some correct parameters of ECAP device and pressing process such as geometry parameters, the pressing speed of the ram, pressing temperature or load displacement curve [1, 2].

3.1. General assumptions in FEM simulation

The FEM is used because it can provide us direct information on the evolution of plastic deformation during the ECAP and enable us to simulate the deformation of materials subjected to single or multi-pass ECAP.

In the FEM simulation, the following assumptions are made [1]:

- First, the material is isotropic and homogeneous.
- Second, the material is elastoplastic with strain-hardening exponent being zero in order to consider the effect of elastic deformation on the morphological change of the extruded billet.
- Third, the system is isothermal.
- Fourth, the von Mises flow rule is used to construct the constitutive relation.
- Fifth, there is no friction between the surface of the material and the die wall due to the use of lubricant in the ECAP.

3.2. FEM formulation of ECAP

It is known that the extruded billet experiences large plastic deformation in the ECAP process. The model based on continuum mechanics needs to allow for arbitrary finite strains.

The elastoplastic behaviour of the billet is described by using a generalization of J_2 flow theory (Prandtl-Reuss equations for isotropic materials with isotropic hardening) to finite strain. In the current configuration x_i, the components of displacement vector and the metric tensors are denoted as u_i and G_{ij}, respectively. In the reference coordinate system x^i, the components of the displacement vector and the metric tensors are denoted as u^i and g_{ij}, respectively. The determinants of G_{ij} and g_{ij} are G and g, respectively. Then the Lagrangian strain tensor η_{ij} is

$$\eta_{ij} = [(u_{i,j} + u_{j,i})/2] + [(u_i^k u_{k,j})/2] \tag{8}$$

where $(\ )_{,j}$ denotes the covariant derivative in the reference frame.

The contravariant components of the Kirchhoff stress τ^{ij} and the Cauchy stress tensor σ^{ij} on the current base vectors are related by

$$\tau^{ij} = \sqrt{G/g} \cdot \sigma^{ij}. \tag{9}$$

Using a generalized J_2 flow theory, the tensor of instantaneous moduli L^{ijkl} relating the stress and strain increments in the constitutive law, $\dot{\tau}^{ij} = L^{ijkl}\dot{\eta}_{kl}$ is given by

$$
L^{ijkl} = \frac{E}{1+\nu}\left[\frac{1}{2}\left(G^{ik}G^{jl} + G^{il}G^{jk}\right) + \frac{\nu}{1-2\nu}G^{ij}G^{kl} - \right.
$$
$$
\left. -\alpha\frac{3}{2}\frac{(E/E_t)-1}{(E/E_t)-(1-2\nu)/3}\frac{s^{ij}s^{kl}}{\sigma_e^2}\right] - \frac{1}{2}\left[G^{ik}\tau^{jl} + G^{jk}\tau^{il} + G^{il}\tau^{jk} + G^{jl}\tau^{ik}\right] \tag{10}
$$

and

$$\alpha = 1 \text{ if } \dot{\sigma}_e \geq 0 \text{ and } \sigma_e = (\sigma_e)_{\max} \qquad \text{or} \qquad \alpha = 0 \text{ if } \dot{\sigma}_e < 0 \text{ and } \sigma_e < (\sigma_e)_{\max}$$

here E and ν are Young's modulus and Poisson's ratio, respectively, $\sigma_e = (3s_{ij}s^{ij}/2)^{1/2}$ is the von Mises stress and $s^{ij} = \tau^{ij} - (G^{ij}G_{kl}\tau^{kl})/3$ is the stress deviator. The tangent modulus E_t is the slope of the uniaxial true stress versus logarithmic strain curve at the stress level $(\sigma_e)_{\max}$. The curve is represented by a piecewise power law

$$\varepsilon = \begin{cases} \dfrac{\sigma}{E} & \text{for } \sigma \leq \sigma_y \\ \dfrac{\sigma_y}{E}\left(\dfrac{\sigma}{\sigma_y}\right)^{1/n} & \text{for } \sigma > \sigma_y \end{cases} \tag{11}$$

where n is the strain-hardening exponent and σ_y is the initial yield stress.

For materials without strain-hardening, the deformation is controlled by the criterion,

$$\sigma_e < \sigma_y \qquad \text{for elastic deformation,} \tag{12}$$

$$\sigma_e = \sigma_y \qquad \text{at yield.} \tag{13}$$

The equivalent plastic strain is defined as

$$\varepsilon^{pl} = \left(\frac{3\varepsilon_{p,ij}\varepsilon_p^{ij}}{2}\right)^{1/2} \tag{14}$$

where $\varepsilon_{p,ij}$ is the plastic strain tensor.

The equations of equilibrium are defined in terms of the principle of virtual work, which gives

$$\int_V \left\{\dot{\tau}^{ij}\delta\eta_{ij} + \tau^{ij}\delta\dot{u}_{,i}^k u_{k,j}\right\} \mathrm{d}V = \int_S \dot{T}^i \delta u_i \, \mathrm{d}S - \left[\int_V \tau^{ij}\delta\eta_{ij} \, \mathrm{d}V - \int_S T^i \delta u_i \, \mathrm{d}S\right] \tag{15}$$

where S and V denote the surface and the volume, respectively, in the reference state and T^i are the contravariant components of the nominal surface tractions. The bracketed terms are included to prevent drifting away from the true equilibrium path.

To simplify the FEM analysis of the ECAP process and increase the computational efficiency, we only analyze two-dimensional problem. Boundary conditions need to be defined to completely determine the deformation behaviour of the billet. Fig. 3 shows the general principle of the ECAP process. The surface of the billet can be divided into six segments (S_{AB}, S_{BC}, S_{CD}, S_{DE}, S_{EF} and S_{FA}), depending on the extrusion process.

Considering the use of lubricant in the ECAP, we assume that the contact between the billet and the surfaces of die and punch is frictionless in the FEM simulation. For the segments S_{BC} and S_{EF}, there are

$$u_2 = u_0, \ \sigma^{12} = 0 \text{ and}$$

$$F = \int_{S_{BC}} \sigma^{22} \, \mathrm{d}S \text{ on the segment } S_{BC} \tag{16}$$

$$\sigma^{ij}n_i n_j = 0 \text{ on the segment } S_{EF} \tag{17}$$

where n_i are the components of the unit normal vector of the segment S_{EF}.

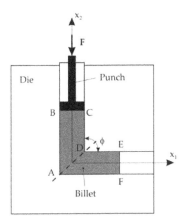

Fig. 3. Schematic diagram of the ECAP process

The boundary conditions for segments S_{AB}, S_{CD}, S_{DE} and S_{FA} involves the inequality equations, which can be expressed as

$$u_1 \geq 0 \text{ and } \sigma^{21} = 0 \text{ on the segment } S_{AB} \tag{18}$$

$$u_1 \leq 0 \text{ and } \sigma^{21} = 0 \text{ on the segment } S_{CD} \tag{19}$$

$$u_i n^i \leq 0 \text{ and } \sigma^{ij} n_i n_j = 0 \quad (i \neq j) \text{ on the segment } S_{DE} \tag{20}$$

$$u_i n^i \geq 0 \text{ and } \sigma^{ij} n_i n_j = 0 \quad (i \neq j) \text{ on the segment } S_{FA} \tag{21}$$

where $n^i = n_i$. The inequality indicates that the billet cannot penetrate onto the surface of the die, while separation of the billet from the die can occur.

4. Used finite element models and simulation parameters

The mechanical response of an elastoplastic billet and the deformation behaviour of ECAP process using the commercial FEM software ADINA were made. It was analyzed influence of shape of the inner and outer corner of die.

Very important factor in FEM simulations of forming processes is to find the suitable combination of the mesh density and suitable choice of element type. In these cases occur large deformations – large displacements and large strains. Furthermore, formed bodies are in contact condition with dies or tools during forming process. It means that chosen type of finite elements must fully satisfy contact and deformation conditions and size of the mesh fineness must secure that there will not be too much distortion in these elements during whole simulation. In FEM simulation three nodes linear plane strain elements were used. All FEM simulations were carried out with the same geometry of workpiece and finite element mesh. The two-dimensional workpiece considered has the dimensions of 10 mm (width) $\times 50$ mm (length) and a unity thickness since a plane strain condition is assumed. The workpiece consisted of $18\,426$ triangular elements.

The deformation behaviour of Al material workpiece during extrusion process of ECAP was simulated and analyzed. The extrusion of the Al material billet was made by assuming isothermal conditions at room temperature ($T = 20°$) and neglecting the heating conditions due to the friction between the workpiece and the die tool. The details around geometry and main parameters of used FEM models are listed in tab. 1.

Table 1. List of the name and basic parameters of used FEM mode

Name of model	ϕ	ψ	Inner corner radius [mm]
m01	90°	0°	–
m02	90°	90°	–
m03	90°	90°	0,1× width
m04	90°	90°	0,2× width
m05	90°	90°	0,3× width
m06	90°	90°	0,4× width
m07	90°	90°	0,5× width
m08	90°	90°	0,6× width
m09	90°	90°	0,7× width
m10	90°	90°	0,8× width
m11	90°	90°	0,9× width
m12	90°	90°	1,0× width

For this purpose 2D models were developed using a plane strain condition. The coefficient of friction between the inside of the die channel and the workpiece was assumed to be zero implying a frictionless condition. A constant pressing speed of 1 mms^{-1} was imposed in each simulation. This constant movement was secured through the displacement function. Workpiece model contained 24 points. Moreover, all points were organized into the four cutting planes and six levels together. These cutting planes and levels can be seen in fig. 4.

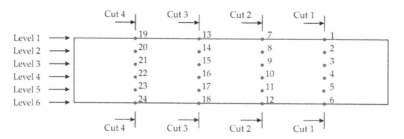

Fig. 4. Analyzed points, cut planes and levels in the workpiece

In general, three combinations of the inner and outer corner shapes are used in practice. Fig. 5 shows these the most widely used variants.

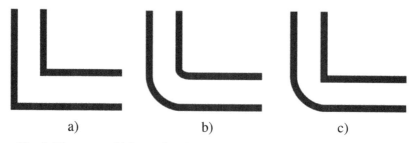

a)   b)   c)

Fig. 5. The most widely used variants of the inner and outer corner shapes

It was necessary to create two different element groups because two kind of material was used. The first material represents material used for the die and plunger. These parts was modeled with the isotropic linear elastic material with the Young's modulus E and the Poisson's ratio ν equal to $210\,000$ MPa and $0{,}3$ respectively. The second material represents workpiece. The material model with isotropic hardening and diagram of true stress strain behaviour was used. The workpiece material used in the calculations was pure Al, which exhibits strain-rate sensitivity of flow stress and strain-hardening behaviour. The stress-strain curve used for the multilinear elastic-plastic material model was obtained from tensile and compression experimental tests. These obtained values were direct input data for this material model. In this material model was considered with the Young's modulus E and the Poisson's ratio ν equal to $72\,000$ MPa and $0{,}3$ respectively.

5. Obtained results

In the first variant both inner and outer corners are sharp. This combination is shown in fig. 6 and provides that the accumulative effective plastic strains are very high in comparison to the other variants. Main disadvantages of this variant are expensiveness of dies and frequent exchanges of dies because inner sharp corner is usually worn-out very rapidly.

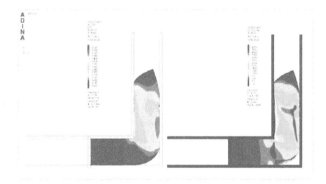

Fig. 6. Distribution of the effective stress and accumulative effective plastic strain in the workpiece during the ECAP process – $t = 25$ [s] – variant with both sharp corners

In the second variant both inner and outer corners have rounded shape. This variant represents fig. 7. Obtained values of the accumulative effective plastic strain are much lower than in the previous variant. In practice despite of this fact the variant with both round corners is the most widely used. The inner round corner has usually very small radius. Main advantage of this variant is that no empty are made in contrast to the first variant.

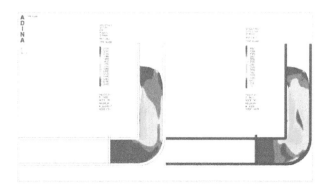

Fig. 7. Distribution of the effective stress and accumulative effective plastic strain in the workpiece during the ECAP process – $t = 40$ [s] – variant with both round corners.

Fig. 8 shows variant where inner corner is sharp and outer corner has round shape. This variant is combination of the first two above-mentioned variants. Acquired values of effective stress and accumulative effective plastic strain are lower than in the first variant but are higher than in the second variant.

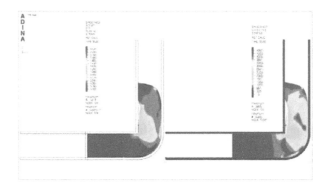

Fig. 8. Distribution of the effective stress and accumulative effective plastic strain in the workpiece during the ECAP process – $t = 30$ [s] – variant with sharp/round corners

There are presented histograms (see fig. 9) for each analyzed point of interest. Points which lies in the vicinity of the inner side of die and in the front of the workpiece especially point 1 gave inaccurate results. This is valid mainly for the first two variants – in both is presence of the sharp inner corner.

Obtained results of the accumulative effective plastic strain presented above in histograms can lead to expected conclusions. If it takes only points situated out of problematic zones (front part of workpiece and points lies in the vicinity of the inner side of die) thus can be noted that the highest values are obtained from the first variant with both sharp corners.

For passing the workpiece through the die it is necessary to know the pressing force because when it is used the high pressing force the tool of ECAP device can be damaged. Fig. 10 shows the pressing forces acting on the workpiece during ECAP process in analyzed variants.

6. Conclusion

It was necessary to analyze behaviour of the accumulative effective plastic strain and effective stress in different region of the aluminium workpiece. Another important data with respect to safe operation mode of pressing device is pressing force acting on the plunger and workpiece, respectively.

FEM analysis of deformation bahaviour during ECAP process showed the tremendous increase of plastic deformation. This fact can be easily seen from the histograms of the accumulative effective plastic strain. From obtained results can be seen that only in the points situated in the edge of workpiece passing over the inner die (top side of workpiece) occur higher straining than in the rest of workpiece. This fact is valid for all carried out simulations.

Both the largest and uniform accumulative effective plastic strain increased with the decrease of the radius of the corner die.

The force required to press the workpiece through the die decreased with the increase in the radius of the corner die. This fact can be easily seen from the courses of pressing forces in the fig. 10. The courses are similar but obtained values are different. The first variant with

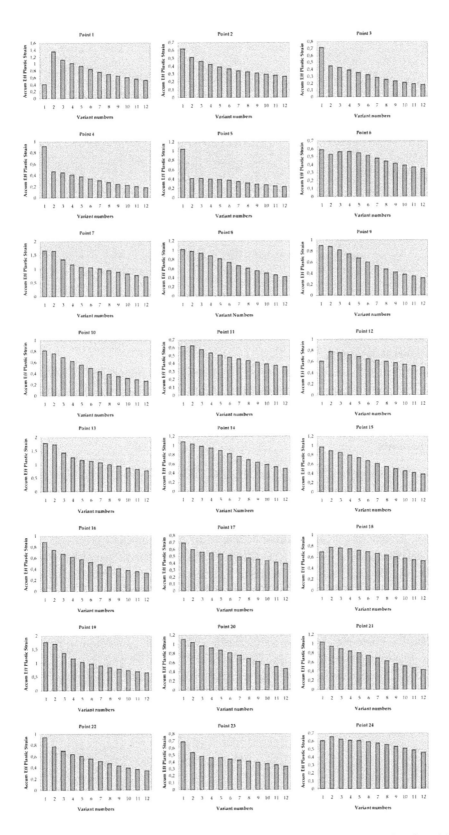

Fig. 9. Comparation of values of accumulative effective plastic strain in all 24 points in m01–m12

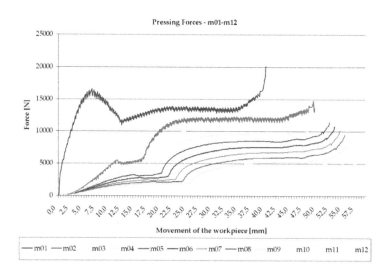

Fig. 10. Pressing forces acting on the workpiece in used variants m01–m12

both sharp corners needs the highest pressing force in the first stage of pressing. This fact fully satisfied the general assumption.

The precision of obtained results can be influenced by contact and its weak formulation. Another fact that can influence the results may be the discretization of solution on the finite elements.

Acknowledgements

The work has been supported by the grant projects KEGA 3/5028/07 and VEGA 1/0125/09.

References

[1] Melicher, R., Numerical simulation of bulk nanostructured materials processed by the method of equal channel angular pressing, Ph. D. thesis, Žilinská univerzita, Strojnícka fakulta, Žilina 2008.

[2] Melicher, R., Handrik, M., Analysis of shape parameters of tool for ECAP technology, Acta Mechanica Slovaca 3-C/2008, Košice, 2008, pp. 273–284. (in Slovak)

[3] Melicher, R., Finite element method simulation of equal channel angular pressing, In Transcom 2009, Žilina, 2009, p. 91–94.

[4] Melicher, R., Finite element analysis of plastic deformation of aluminium specimen by ECAP process. In Nekonvenčné technológie 2008, 21/235, p. 1–12. (in Slovak)

[5] Melicher, R., Mechanical modeling of severe plastic deformation by ECAP and ECAR technology, Písomná časť dizertačnej práce, Žilinská univerzita, Strojnícka fakulta, Žilina 2007. (in Slovak)

13

Analytical and numerical investigation of trolleybus vertical dynamics on an artificial test track

P. Polach[a,*], M. Hajžman[a], J. Soukup[b], J. Volek[b]

[a] *Section of Materials and Mechanical Engineering Research, ŠKODA VÝZKUM, s. r. o., Tylova 1/57, 316 00 Plzeň, Czech Republic*

[b] *Department of Mechanics and Machines, Faculty of Production Technology and Management, University of J. E. Purkyně in Ústí nad Labem, Na Okraji 1001, 400 96 Ústí nad Labem, Czech Republic*

Abstract

Two virtual models of the ŠKODA 21 Tr low-floor trolleybus intended for the investigation of vertical dynamic properties during the simulation of driving on an uneven road surface are presented in the article. In order to solve analytically vertical vibrations, the trolleybus model formed by the system of four rigid bodies with seven degrees of freedom coupled by spring-damper elements is used. The influence of the asymmetry of a sprung mass, a linear viscous damping and a general kinematic excitation of wheels are incorporated in the model. The analytical approach to solving the ŠKODA 21 Tr low-floor trolleybus model vibrations is a suitable complement of the model based on a numerical solution. Vertical vibrations are numerically solved on the trolleybus multibody model created in the *alaska* simulation tool. Both virtual trolleybus models are used for the simulations of driving on the track composed of vertical obstacles. Conclusion concerning the effects of the usage of the linear and the nonlinear spring-damper elements characteristics are also given.

Keywords: vehicle dynamics, trolleybus, analytical solution, numerical simulation, multibody model

1. Introduction

Computational models of vehicles, which are used in vehicle dynamics tasks, can be of a various complexity and therefore it is efficient to have a variety of models with respect to their application. General approaches to the vehicle modelling and their reviews can be found in [1] and [12]. The advantage of simple models (concerning kinematic structure and number of degrees of freedom) is mainly the shorter computational time of particular analyses. They can be used for a sensitivity analysis, optimization [15], parameters identification etc. The problems of interaction [5] are also studied very often with this sort of models. On the other hand more complex multibody models [4, 6, 10] can be used for detailed analyses and for the investigation of the chosen structural elements behaviour. The most of published works are based on numerical simulations with the created vehicle models and the analytical methods, which can bring faster and more accurate analyses, are omitted.

In connection with previous contributions to the investigation of vertical vibration of vehicles under various conditions [2, 3, 7, 8, 9, 10, 11, 13, 14, 16, 17] this article deals with the analytical and numerical solutions of vertical vibration of the empty ŠKODA 21 Tr low-floor trolleybus (see fig. 1) models.

*Corresponding author. e-mail: pavel.polach@skodavyzkum.cz.

Fig. 1. The ŠKODA 21 Tr low-floor trolleybus

The usage of the simplified analytical model for the dynamic analysis of road vehicles is justified because of the transparency of a mathematical model, easier implementation and the possibility of a better understanding of the mechanical system behaviour. Analytical solution enables to put the monitored quantities (displacements, velocities, accelerations) in the form of continuous function of time (enabling the analytical performing of derivative and integration) in contradiction to the discrete form of those quantities obtained by means of the numerical solution. Relations for the calculation of the monitored quantities in the whole investigated period of time are obtained during the analytical solving of the equations of motion. The numerical method requires to solve the equations of motion for each integrating step of the investigated period of time. In order to solve numerically the vertical vibration the trolleybus multibody model created in the *alaska* simulation tool is used.

The main differences between the models are in the consideration of linear characteristics of spring-damper elements and in the impossibility of including the bounce of the tire from the road surface in the analytical model.

2. Analytical solution

For the analytical solution the trolleybus model is considered to be a system of four rigid bodies with seven degrees of freedom, coupled by spring and dissipative elements (see fig. 2), taking into account a linear viscous damping and the influence of asymmetry (e.g. mass distribution) and with general kinematic excitation of the individual wheels. The rigid bodies correspond to the sprung mass (trolleybus body) and the unsprung masses – the rear axle (including wheels) and the front half axles (including wheels). Spring and dissipative elements model the tire-road surface contact (two front and four rear wheels), the air springs in the axles suspension (two front and four rear ones) and the hydraulic shock absorbers in the axles suspension (two front and four rear ones). As it was already mentioned, characteristics of the spring and dissipative elements are supposed to be linear.

In general case, for the considered trolleybus model it is possible to put the equations of motion in the matrix form (see [14])

$$\mathbf{M} \cdot \ddot{\mathbf{q}}(t) + \mathbf{B} \cdot \dot{\mathbf{q}}(t) + \mathbf{K} \cdot \mathbf{q}(t) = \mathbf{f}(t), \tag{1}$$

where $\mathbf{q}(t) = [\varphi, \theta, z, z_1, z_2, z_3, \varphi_3]^T$, $\dot{\mathbf{q}}(t)$, $\ddot{\mathbf{q}}(t)$ are the vectors of the generalized coordinates (φ is the angular displacement of the trolleybus body – sprung mass – around the longitudinal x-axis, θ is the angular displacement of the trolleybus body around the lateral y-axis, z is the

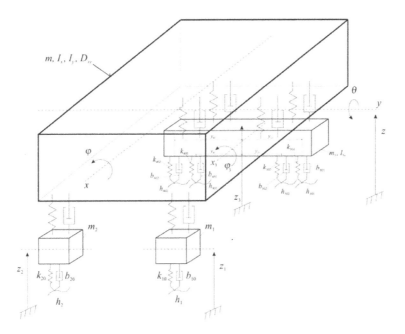

Fig. 2. The scheme of the trolleybus analytical model

vertical displacement of the trolleybus body, z_1 is the vertical displacement of the left front half axle, z_2 is the vertical displacement of the right front half axle, z_3 is the vertical displacement of the rear axle, φ_3 is the angular displacement of the rear axle around the longitudinal axis) and their derivatives with respect to the time, $\mathbf{M}$ is the mass matrix (diagonal elements: I_x – the moment of inertia with respect to x axis of the trolleybus body, I_y – the moment of inertia with respect to y axis of the trolleybus body, m – the mass of the trolleybus body, m_1 – the mass of the left front half axle, m_2 – the mass of the right front half axle, m_3 – the mass of the rear axle, I_{3x} – the moment of inertia with respect to x axis of the rear axle; other elements in case of a symmetric distribution of the sprung mass $m_{ik} = m_{ki} = 0$ for $i \neq k$ and for $i = 1, 2, \ldots, 7$, $k = 1, 2, \ldots, 7$; in case of an asymmetric distribution of the sprung mass $m_{12} = m_{21} = -D_{xy}$, where D_{xy} is the product of inertia with respect to x and y axes of the trolleybus body), $\mathbf{B}$ is the damping matrix, $\mathbf{K}$ is the stiffness matrix (due to the generality all the elements are considered not to be zero), $\mathbf{f}(t) = [f_i(t)]^T$ for $i = 1, 2, \ldots, 7$, is the vector of the generalized forces (kinematic excitation function) [14].

After dividing the individual equations by the respective diagonal element of mass matrix $\mathbf{M}$ (suitable mathematical adjustment due to the solution procedure) and after the Laplace integral transform for the zero initial conditions, i.e. in time $t = 0$ $\mathbf{q}(0) = 0$ and $\dot{\mathbf{q}}(0) = 0$, the system of differential equations is transformed to the system of algebraic equations

$$\mathbf{S} \cdot \bar{\mathbf{q}}(s) = \bar{\mathbf{f}}(s), \tag{2}$$

where s is the parameter of transform, $\bar{\mathbf{q}}(s)$ are the images of the vector of generalized co-ordinates $\mathbf{q}(t)$ and $\bar{\mathbf{f}}(s)$ are the images of the vector of generalized forces $\mathbf{f}(t)$ divided by the respective diagonal element of mass matrix $\mathbf{M}$.

It holds for the elements of the matrix $\mathbf{S}$:
$$a_{ij} = s^2 + \beta_{ij} \cdot s + \kappa_{ij}, \qquad \text{for } i = j,$$
$$a_{ij} = \delta_{ij} \cdot s^2 + \beta_{ij} \cdot s + \kappa_{ij}, \quad \text{for } i \neq j \text{ and for } i = 1, 2 \text{ and } j = 1, 2,$$

$$a_{ij} = \beta_{ij} \cdot s + \kappa_{ij}, \quad \text{for } i \neq j \text{ and for } i = 1, 2 \text{ and } j = 3, 4, \ldots, 7,$$
$$\text{for } i \neq j \text{ and for } i = 3, 4, \ldots, 7 \text{ and } j = 1, 2, \ldots, 7,$$

where $\kappa_{ij} = \frac{k_{ij}}{m_{ii}}$ and $\beta_{ij} = \frac{b_{ij}}{m_{ii}}$ can be calculated from the original elements of stiffness matrix $\mathbf{K}$ and damping matrix $\mathbf{B}$ after division of the equations by the diagonal elements of mass matrix $\mathbf{M}$, $\delta_{12} = -\frac{D_{xy}}{I_x}$ and $\delta_{21} = -\frac{D_{xy}}{I_y}$ are the elements respecting the influence of asymmetric distribution of the mass of the trolleybus body.

For solving the system of algebraic equations (2), i.e. for determining the images of generalized coordinates $\bar{q}_j(s)$, $j = 1, 2, \ldots, 7$, it is possible, due to a small number of equations, to apply the Cramer rule

$$\bar{q}_j(s) = \frac{D_j(s)}{D(s)}, \tag{3}$$

where $D(s)$ is the determinant of matrix $\mathbf{S}$ and $D_j(s)$ is the determinant which originates from determinant $D(s)$ by replacing the j-th column of elements a_{ij} ($i = 1, 2, \ldots, 7$) of determinant $D(s)$ with the column of right sides of the system of linear algebraic equations (2), i.e. with the elements of vector $\bar{\mathbf{f}}(s)$. This method is suitable regarding the process of further solving, i.e. obtaining the vector of generalized coordinates $\mathbf{q}(t)$ by the inverse transform.

For the expansion of determinant $D(s)$ of matrix $\mathbf{S}$ into the form of the polynomial

$$D(s) = \sum_{i=0}^{n=14} A_{n-i} \cdot s^{n-i}, \tag{4}$$

where the polynomial degree n is given by the double of degrees of freedom of the mechanical system (i.e. $n = 14$), it is necessary to determine coefficients A_{n-i} for $i = 1, 2, \ldots, n$ ($A_n = 1$). This operation can be carried out by means of symbolic calculations using the specialized mathematical software.

By evaluating determinant $D_j(s)$ the relation for images $\bar{q}_j(s)$ of function $q_j(t)$ is obtained

$$\bar{q}_j(s) = \sum_{i=1}^{7} (-1)^{j+i} \cdot \bar{f}_i(s) \cdot \frac{D_{ji}(s)}{D(s)}, \quad \text{for } j = 1, 2, \ldots, 7, \tag{5}$$

where determinant $D_{ji}(s)$ is a subdeterminant of order $n/2 - 1$ of determinant $D(s)$ corresponding to element a_{ij} of matrix $\mathbf{S}$.

The polynomial corresponding to subdeterminant $D_{ji}(s)$ is determined using the same algorithm as the polynomial (4) if the value of element a_{ij} is changed to $a_{ij} = (-1)^{i+j}$, the other elements $a_{i,j\neq r}$ ($r = 1, 2, \ldots, n$) in the row i are set to zero $a_{i,j\neq r} = 0$ ($r = 1, 2, \ldots, n$) and the other elements $a_{i\neq k,j}$ ($k = 1, 2, \ldots, n$) in the column j are set to zero $a_{i\neq k,j} = 0$ ($k = 1, 2, \ldots, n$), while the values of the other elements $a_{i\neq k,j\neq r}$ ($k = 1, 2, \ldots, n$, $r = 1, 2, \ldots, n$) of determinant $D(s)$ do not change

$$D_{ji}(s) = \sum_{r=0}^{m} d_{ji,m-r} \cdot s^{m-r}, \quad \text{for } j = 1, 2, \ldots, n/2, \; i = 1, 2, \ldots, n/2, \tag{6}$$

where $m = n - 1$ for $i = j$,
                   for $i \neq j$ and for $i = 1, 2$ and $j = 1, 2$,
and $m = n - 2$ for $i \neq j$ and for $i = 1, 2$ and $j = 3, 4, \ldots, 7$,
                   for $i \neq j$ and for $i = 3, 4, \ldots, 7$ and $j = 1, 2, \ldots, 7$,
and coefficients $d_{ji,m-r}$ for $r = 0, 1, \ldots, m$ can be determined using the symbolic calculations.

In order to determine the original $q_j(t)$ of corresponding image $\bar{q}_j(s)$ it is suitable the relation (5) to be transformed to the form of convolution. That is why it is necessary to calculate the zero points s_k of the polynomial of determinant $D(s)$ [14, 18]. In the given case the zero points s_k are supposed to be in the form of the complex conjugate numbers $s_k = \mathrm{Re}\{s_k\} + \mathrm{i} \cdot \mathrm{Im}\{s_k\}$ and $s_{k+1} = \mathrm{Re}\{s_k\} - \mathrm{i} \cdot \mathrm{Im}\{s_k\}$, for $k = 1, 3, \ldots, n-1$ (i is the imaginary unit). By calculating zero points of the polynomial (4) it is possible, using the product of root factors, to put the polynomial in the form of the product of the quadratic polynomials

$$[s - (\mathrm{Re}\{s_i\} + \mathrm{i} \cdot \mathrm{Im}\{s_i\})] \cdot [s - (\mathrm{Re}\{s_i\} - \mathrm{i} \cdot \mathrm{Im}\{s_i\})] = s^2 + p_i \cdot s + r_i, \text{ for } i = 1, 2, \ldots, n/2, \tag{7}$$

where $r_i = (\mathrm{Re}\{s_i\})^2 + (\mathrm{Im}\{s_i\})^2$ and $p_i = -2 \cdot \mathrm{Re}\{s_i\}$.

According to (4) it yields

$$D(s) = \sum_{k=0}^{n=14} A_{n-k} \cdot s^{n-k} = \prod_{i=1}^{n/2} \left(s^2 + p_i \cdot s + r_i\right). \tag{8}$$

Then it is possible to transfer the ratio of the determinants in equation (5) to the sum of partial fractions (supposing simple roots) in the form [14]

$$\frac{D_{ji}(s)}{D(s)} = \frac{\sum\limits_{r=0}^{m} d_{ji,m-r} \cdot s^{m-r}}{\prod\limits_{k=1}^{n/2} \left(s^2 + p_k \cdot s + r_k\right)} = \frac{\sum\limits_{r=1}^{n/2} \left[(K_{ji,r} \cdot s + L_{ji,r}) \cdot \prod\limits_{\substack{k=1 \\ k \neq r}}^{n/2} \left(s^2 + p_k \cdot s + r_k\right) \right]}{\prod\limits_{k=1}^{n/2} \left(s^2 + p_k \cdot s + r_k\right)}, \tag{9}$$

where constants $K_{ji,r}$ and $L_{ji,r}$ for $j = 1, 2, \ldots, n/2$, $i = 1, 2, \ldots, n/2$, $r = 1, 2, \ldots, n/2$, can be determined from the condition of the coefficients equality at identical powers of parameter s in numerators of fraction on both sides of equation (9).

The condition of the numerators equality (9) can be expressed by the relation

$$\sum_{r=0}^{m} d_{ji,m-r} \cdot s^{m-r} = \sum_{r=1}^{n/2} \left[(K_{ji,r} \cdot s + L_{ji,r}) \cdot \frac{\prod\limits_{k=1}^{n/2} \left(s^2 + p_k \cdot s + r_k\right)}{s^2 + p_r \cdot s + r_r} \right], \tag{10}$$

where m ($m = n - 1$ or $m = n - 2$) is the order of the polynomial of determinant $D_{ji}(s)$, j is the designation of the component of vector of the images of generalized coordinates $\bar{q}_j(s)$ ($j = 1, 2, \ldots, n/2$) and i is the designation of the component of vector of the images of the generalized forces (divided by the respective diagonal element of mass matrix $\mathbf{M}$) ($i = 1, 2, \ldots, n/2$).

As the denominator of fraction on the right side of equation (10) is the divisor of the numerator (i.e. division remainder is equal zero), this fraction can be, regarding the relation (8), modified to the form [14, 18]

$$\frac{\prod\limits_{k=1}^{n/2} \left(s^2 + p_k \cdot s + r_k\right)}{s^2 + p_i \cdot s + r_i} = \frac{\sum\limits_{k=0}^{n} A_{n-k} \cdot s^{n-k}}{s^2 + p_i \cdot s + r_i} = \sum_{k=2}^{n} t_{i,n-k} \cdot s^{n-k}, \text{ for } i = 1, 2, \ldots, n/2, \tag{11}$$

where $t_{i,n-1} = 0$,

$\quad t_{i,n-2} = A_n = 1$,

$\quad t_{i,n-k} = A_{n-k+2} - p_i \cdot t_{i,n-k+1} - r_i \cdot t_{i,n-k+2}$, for $k = 3, 4, \ldots, n$, $i = 1, 2, \ldots, n/2$.

Then equation (10) can be put in the form

$$\sum_{r=0}^{m} d_{ji,m-r} \cdot s^{m-r} = \sum_{r=1}^{n/2} \left[(K_{ji,r} \cdot s + L_{ji,r}) \cdot \sum_{k=2}^{n} t_{r,n-k} \cdot s^{n-k} \right], \tag{12}$$

for $j = 1, 2, \ldots, n/2$, $i = 1, 2, \ldots, n/2$.

After performing the multiplication of the polynomials on the right side of equation (12) and comparing the coefficients of identical powers of parameter s on both sides of equation (12) a system of n algebraic equations for unknown coefficients $K_{ji,r}$ and $L_{ji,r}$ for $j = 1, 2, \ldots, n/2$, $i = 1, 2, \ldots, n/2$ and $r = 1, 2, \ldots, n/2$ is obtained

$$\sum_{r=1}^{n/2} (K_{ji,r} \cdot t_{r,n-k-1} + L_{ji,r} \cdot t_{r,n-k}) = d_{ji,n-k}, \text{ for } k = 1, 2, \ldots, n, \tag{13}$$

where for $k = n$ coefficients $t_{r,n-k-1} = t_{r,-1} = 0$, $r = 1, 2, \ldots, n/2$, represent the division remainders of equation (11) – see [14] or [18].

By analytical solving the system of algebraic equations (13) – see [14] – unknown coefficients $K_{ji,r}$ and $L_{ji,r}$ are determined and equation (9) (using relation (10)) can be put in the form

$$\frac{D_{ji}(s)}{D(s)} = \sum_{k=1}^{n/2} \frac{K_{ji,k} \cdot s + L_{ji,k}}{s^2 + p_k \cdot s + r_k}, \text{ for } j = 1, 2, \ldots, n/2, i = 1, 2, \ldots, n/2. \tag{14}$$

By means of this relation equation (5) for the calculation of the image of generalized coordinates $\bar{q}_j(s)$, for $j = 1, 2, \ldots, n/2$, can be modified into the form

$$\bar{q}_j(s) = \sum_{i=1}^{n/2} (-1)^{j+i} \cdot \bar{f}_i(s) \cdot \sum_{k=1}^{n/2} \frac{K_{ji,k} \cdot s + L_{ji,k}}{s^2 + p_k \cdot s + r_k}. \tag{15}$$

The denominator of the fraction of equation (15) can be modified to the form

$$s^2 + p_k \cdot s + r_k = (s + \beta_k)^2 + \Omega_k^2, \text{ for } k = 1, 2, \ldots, n/2, \tag{16}$$

where $\Omega_k^2 = \omega_k^2 - \beta_k^2$ is the damped natural frequency, $\omega_k^2 = r_k$ is the undamped natural frequency and $\beta_k = -\frac{p_k}{2}$ is the coefficient of linear viscous damping.

Using formula (16) equation (15) can be written in the form

$$\bar{q}_j(s) = \sum_{i=1}^{n/2} (-1)^{j+i} \cdot \bar{f}_i(s) \cdot \sum_{k=1}^{n/2} \frac{K_{ji,k} \cdot s + L_{ji,k}}{(s + \beta_k)^2 + \Omega_k^2}, \text{ for } j = 1, 2, \ldots, 7. \tag{17}$$

Further modification may result in the final relation for image $\bar{q}_j(s)$ for $j = 1, 2, \ldots, 7$

$$\bar{q}_j(s) = \sum_{i=1}^{n/2} (-1)^{j+i} \cdot \bar{f}_i(s) \cdot \sum_{k=1}^{n/2} \left[K_{ji,k} \cdot \frac{s + \beta_k}{(s + \beta_k)^2 + \Omega_k^2} + \right. \tag{18}$$

$$\left. + \frac{L_{ji,k} - \beta_k \cdot K_{ji,k}}{\Omega_k} \cdot \frac{\Omega_k}{(s + \beta_k)^2 + \Omega_k^2} \right].$$

After the inverse transform of the relation (18), the function of generalized coordinate $q_j(t)$, for $j = 1, 2, \ldots, 7$, in the form of the sum of convolution integrals is obtained

$$q_j(t) = \sum_{i=1}^{n/2} (-1)^{j+i} \cdot \sum_{k=1}^{n/2} \left\{ K_{ji,k} \cdot \int_0^t \frac{f_i(\tau)}{m_{ii}} \cdot e^{-\beta_k \cdot (t-\tau)} \cdot \cos\left[\Omega_k \cdot (t-\tau)\right] \cdot d\tau + \right.$$
$$\left. + \frac{L_{ji,k} - \beta_k \cdot K_{ji,k}}{\Omega_k} \cdot \int_0^t \frac{f_i(\tau)}{m_{ii}} \cdot e^{-\beta_k \cdot (t-\tau)} \cdot \sin\left[\Omega_k \cdot (t-\tau)\right] \cdot d\tau \right\}, \qquad (19)$$

where the elements of the excitation force vector are generally given by the relations

$$\begin{aligned}
f_1(t) &= 0, \\
f_2(t) &= 0, \\
f_3(t) &= 0, \\
f_4(t) &= k_{10} \cdot h_1(t) + b_{10} \cdot \dot{h}_1(t), \qquad\qquad (20)\\
f_5(t) &= k_{20} \cdot h_2(t) + b_{20} \cdot \dot{h}_2(t), \\
f_6(t) &= k_{301} \cdot h_{31}(t) + k_{302} \cdot h_{32}(t) + k_{401} \cdot h_{41}(t) + k_{402} \cdot h_{42}(t) + \\
&\quad + b_{301} \cdot \dot{h}_{31}(t) + b_{302} \cdot \dot{h}_{32}(t) + b_{401} \cdot \dot{h}_{41}(t) + b_{402} \cdot \dot{h}_{42}(t), \\
f_7(t) &= -k_{301} \cdot h_{31}(t) \cdot y_{31} - k_{302} \cdot h_{32}(t) \cdot y_{32} + k_{401} \cdot h_{41}(t) \cdot y_{41} + k_{402} \cdot h_{42}(t) \cdot y_{42} + \\
&\quad - b_{301} \cdot \dot{h}_{31}(t) \cdot y_{31} - b_{302} \cdot \dot{h}_{32}(t) \cdot y_{32} + b_{401} \cdot \dot{h}_{41}(t) \cdot y_{41} + b_{402} \cdot \dot{h}_{42}(t) \cdot y_{42},
\end{aligned}$$

where $h_1(t)$, $h_2(t)$, $h_{31}(t)$, $h_{32}(t)$, $h_{41}(t)$, $h_{42}(t)$ are the functions describing the shape of the road surface unevenness under the tires, $\dot{h}_1(t)$, $\dot{h}_2(t)$, $\dot{h}_{31}(t)$, $\dot{h}_{32}(t)$, $\dot{h}_{41}(t)$, $\dot{h}_{42}(t)$ are the first-order derivatives of the functions $h_1(t)$, $h_2(t)$, $h_{31}(t)$, $h_{32}(t)$, $h_{41}(t)$, $h_{42}(t)$, k_{10}, k_{20}, k_{301}, k_{302}, k_{401}, k_{402} are the (linear) radial stiffnesses of the tires, b_{10}, b_{20}, b_{301}, b_{302}, b_{401}, b_{402} are the (linear) coefficients of radial damping of the tires and y_{31}, y_{32}, y_{41}, y_{42} are the lateral coordinates of the centres of mass of the rear tires. Subscripts 1 and 10 belong to the left front tire, subscripts 2 and 20 to the right front tire, subscripts 31 and 301 to the left rear outside tire, subscripts 32 and 302 to the left rear inside tire, subscripts 41 and 401 to the right rear inside tire and subscripts 42 and 402 to the right rear outside tire (see fig. 2). See [13] or [14] for more details.

3. Numerical model

The multibody model of the ŠKODA 21 Tr low-floor trolleybus is created in the *alaska 2.3* simulation tool. As it is the first comparison of the results of the simulations performed with the analytical model and with the numerical model there is not utilized the most complex multibody model (formed by 35 rigid bodies and two superelements mutually coupled by 52 joints – e.g. [7, 8]; see fig. 3), but the simplest multibody model created in the *alaska 2.3* simulation tool (e.g. [7]; see fig. 4). The multibody model of the trolleybus is formed by 29 rigid bodies mutually coupled by 29 kinematic joints. The rigid bodies correspond generally to the vehicle individual structural parts. The number of degrees of freedom in kinematic joints is 47.

The rigid bodies are defined by inertia properties (mass, centre of mass coordinates and moments of inertia). The air springs and the hydraulic shock absorbers in the axles suspension and the bushings in the places of mounting some trolleybus structural parts are modelled by connecting the corresponding bodies by nonlinear spring-damper elements. When simulating driving on the uneven road surface the contact point model of tires is used in the multibody model; radial stiffness and radial damping properties of the tires are modelled by nonlinear spring-damper elements considering the possibility of bounce of the tire from the road surface [3, 8, 9].

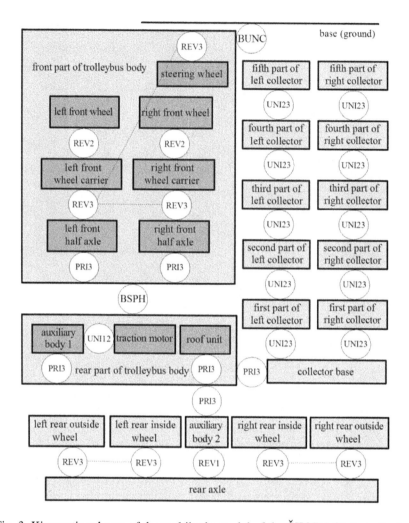

Fig. 3. Kinematic scheme of the multibody model of the ŠKODA 21 Tr trolleybus

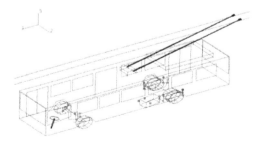

Fig. 4. Visualization of the ŠKODA 21 Tr trolleybus multibody model in the *alaska 2.3* simulation tool

The kinematic scheme of the multibody model of the ŠKODA 21 Tr low-floor trolleybus is shown in fig. 3, where circles represent kinematic joints (BUNC – unconstrained, BSPH – spherical, UNI12 – universal around axes 1 and 2, UNI23 – universal around axes 2 and 3, PRI3 – prismatic in axis 3 direction, REV1 – revolute around axis 1, REV2 – revolute around axis 2, REV3 – revolute around axis 3; axes of the coordinate system are considered according to fig. 4) and quadrangles represent rigid bodies.

4. Simulations results

In order to illustrate the vertical dynamic response calculated by means of the analytical approach and numerical simulation with the trolleybus virtual models, the driving on the artificially created test track according to the ŠKODA VÝZKUM methodology was chosen (e.g. [2, 3, 7, 8, 9, 10]). The test track consisted of three standardized artificial obstacles (in compliance with the Czech Standard ČSN 30 0560 Obstacle II: $h = 60$ mm, $R = 551$ mm, $d = 500$ mm) spaced out on the smooth road surface 20 meters one after another. The first obstacle was run over only with right wheels, the second one with both and the third one only with left wheels (see fig. 5). Results of the drive at the trolleybus models' speed 40 km/h (the usual trolleybus speed according to the ŠKODA VÝZKUM methodology at the driving on the artificially created test track) are shown.

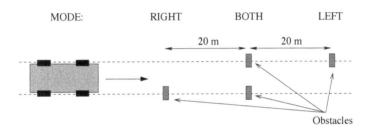

Fig. 5. Visualization of the ŠKODA 21 Tr trolleybus multibody model in the *alaska 2.3* simulation tool

In the course of the test drives simulations time histories of the vertical displacements (see fig. 2) of the trolleybus body z (fig. 6), the left front half axle z_1 (fig. 7), the right front half axle z_2 (fig. 8) and the rear axle z_3 (fig. 9) were monitored (among others).

On the basis of the monitored quantities given in figs 6 to 9 it is generally possible to say that in the results obtained using nonlinear numerical model extreme values of the time histories of the vertical displacements are higher, the decay of dynamic responses of the unsprung masses to the kinematic excitation of the wheels is slower and the decay of dynamic response of the sprung mass to the kinematic excitation of the wheels is faster than in the results obtained using the linear model. But the results are not as different as it was supposed.

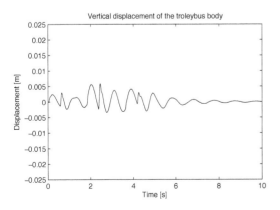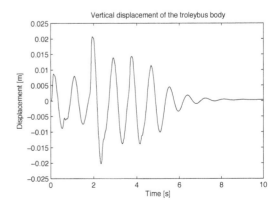

Fig. 6. Time histories of the vertical displacement of the trolleybus body: of the linear model of the trolleybus – left, of the nonlinear model of the trolleybus – right

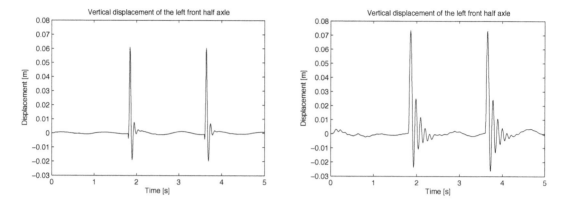

Fig. 7. Time histories of the vertical displacement of the left front half axle: of the linear model of the trolleybus – left, of the nonlinear model of the trolleybus – right

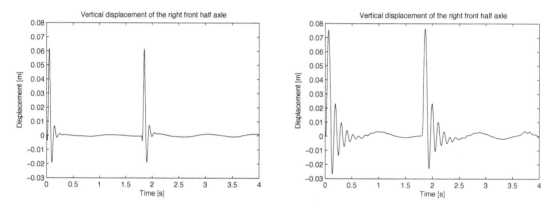

Fig. 8. Time histories of the vertical displacement of the right front half axle: of the linear model of the trolleybus – left, of the nonlinear model of the trolleybus – right

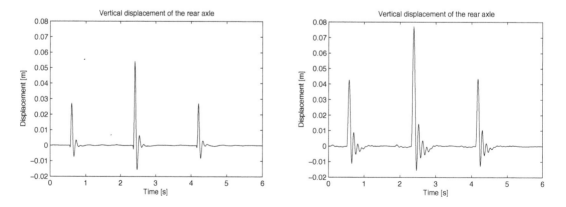

Fig. 9. Time histories of the vertical displacement of the rear axle: of the linear model of the trolleybus – left, of the nonlinear model of the trolleybus – right

On the basis of the test simulations with the trolleybus model created in the *alaska* simulation tool the following facts were stated: the consideration of the only linear force-velocity characteristics of the shock absorbers in the linear model of the trolleybus has the greatest influence on the extreme values of the vertical displacement of the sprung mass, the consideration of the only linear radial stiffness of the tires in the linear model has the greatest influence on the extreme values of the displacements of the unsprung masses, the consideration of the only linear force-deformation characteristics of the air springs and (at the trolleybus models speed 40 km/h) the nonconsideration of the possibility of the tire bounce from the road surface in the linear model have the greatest influence on the decay of the dynamic response of the trolleybus body and the consideration of the only linear radial stiffness of the tires in the linear model has the greatest influence on the decay of the responses of the unsprung masses.

5. Conclusion

Two virtual models of the ŠKODA 21 Tr low-floor trolleybus intended for the investigation of vertical dynamic properties during the simulation of driving along the uneven road surface are presented in the article. The first model and its dynamic response are based on the analytical solution, the second one is based on the multibody modelling and the numerical simulations. Both models have various advantages and together they are complex tools for the vehicle vertical dynamics investigation. The linear analytical model is suitable for the fast and accurate analysis and can be employed mainly for the optimization or the control tasks. The complex multibody model can be used in further steps for a detailed manoeuvre analysis and a particular structural elements evaluation. Differences between the results obtained using both models are discussed.

Acknowledgements

The article has originated in the framework of solving the Research Plan of the Ministry of Education, Youth and Sports of the Czech Republic MSM4771868401 and the No. 101/05/2371 project of the Czech Science Foundation.

References

[1] Blundell, M., Harty, D., The Multibody Systems Approach to Vehicle Dynamics, Elsevier, Oxford, 2004.

[2] Hajžman, M., Polach, P., Optimization Methodology of the Hydraulic Shock Absorbers Parameters in Trolleybus Multibody Models on the Basis of Trolleybus Dynamic Response Experimental Measurement, Proceedings of the International Scientific Conference held on the occasion 55th anniversary of founding the Faculty of Mechanical Engineering of the VSB – Technical University of Ostrava, Session 8 – Applied Mechanics, Ostrava, VSB – TU of Ostrava,, 2005, pp. 117–122.

[3] Hajžman, M., Polach, P., Lukeš, V., Utilization of the trolleybus multibody modelling for the simulations of driving along a virtual uneven road surface, Proceedings of the 16th International Conference on Computer Methods in Mechanics CMM-2005, Czestochowa, Polish Academy of Sciences – Department of Technical Sciences, 2005, CD-ROM.

[4] Hegazy, S., Rahnejat, H., Hussain, K., Multi-Body Dynamics in Full-Vehicle Handling Analysis under Transient Manoeuvre, Vehicle System Dynamics 34 (1) (2000) 1–24.

[5] Hou, K., Kalousek, J., Dong, R., A dynamic model for an asymmetrical vehicle/track system, Journal of Sound and Vibration 267 (3) (2003) 591–604.

[6] Mousseau, C. W., Laursen, T. A., Lidberg, M., Tailor, R. L., Vehicle dynamics simulations with coupled multibody and finite element models, Finite Elements in Analysis and Design 31 (4) (1999) 295–315.

[7] Polach, P., Hajžman, M., Various Approaches to the Low-floor Trolleybus Multibody Models Generating and Evaluation of Their Influence on the Simulation Results, Proceedings of the ECCOMAS Thematic Conference Multibody Dynamics 2005 on Advances in Computational Multibody Dynamics, Madrid, Universidad Politécnica de Madrid, 2005, CD-ROM.

[8] Polach, P., Hajžman, M., Multibody Simulations of Trolleybus Vertical Dynamics and Influences of Various Model Parameters, Proceedings of The Third Asian Conference on Multibody Dynamics 2006, Tokyo, The Japan Society of Mechanical Engineers, 2006, CD-ROM.

[9] Polach, P., Hajžman, M., Multibody simulations of trolleybus vertical dynamics and influences of tire radial characteristics, Proceedings of The 12th World Congress in Mechanism and Machine Science, Besançon, Comité Français pour la Promotion de la Science des Mécanismes et des Machines, 2007, Vol. 4, pp. 42–47.

[10] Polach, P., Hajžman, M., The Investigation of Trolleybus Vertical Dynamics Using an Advanced Multibody Model, Proceedings of the 6th International Conference Dynamics of Rigid and Deformable Bodies 2008, Ústí nad Labem, Faculty of Production Technology and Management, University of J. E. Purkyně in Ústí nad Labem, 2008, pp. 161–170.

[11] Polach, P., Hajžman, M., Volek, J., Soukup, J., Analytical and Numerical Investigation of Vertical Dynamic Response of the Trolleybus Virtual Model, Proceedings of the 9th Conference on Dynamical Systems – Theory and Applications DSTA 2007, Łódź, Department of Automatics and Biomechanics of the Technical University of Łódź, 2007, Vol. 1, pp. 371–378.

[12] Schiehlen, W. (ed.), Dynamical analysis of vehicle systems, Springer, Wien, 2007.

[13] Soukup, J., Volek, J., Investigation of Vertical Vibrations of the ŠKODA 21 Tr Low-floor Trolleybus Model – III, Proceedings of the 4th International Conference Dynamics of Rigid and Deformable Bodies 2006, Ústí nad Labem, University of J. E. Purkyně in Ústí nad Labem, 2006, pp. 191–212. (in Czech)

[14] Soukup, J., Volek, J., Skočilas, J., Skočilasová, B., Segĺa, Š., Šimsová, J., Vibrations of mechanical systems – vehicles, Analysis of influence of the asymmetry, Acta Universitatis Purkynianae 138, Studia Mechanica, University of J. E. Purkyně in Ústí nad Labem, Ústí nad Labem, 2008. (in Czech)

[15] Sun, L., Cai, X., Yang, J., Genetic algorithm-based optimum vehicle suspension design using minimum dynamic pavement load as a design criterion, Journal of Sound and Vibration 301 (1–2) (2007) 18–27.

[16] Volek, J., Soukup, J., Polach, P., Hajžman, M., Investigation of Vertical Vibrations of the ŠKODA 21 Tr Low-floor Trolleybus Model, Proceedings of the National Colloquium with International Participation Dynamics of Machines, Prague, Institute of Thermomechanics AS CR, 2006, pp. 169–174. (in Czech)

[17] Volek, J., Soukup, J., Polach, P., Hajžman, M., Investigation of Vertical Vibrations of the ŠKODA 21 Tr Low-floor Trolleybus Model – II, Proceedings of the National Conference with International Participation Engineering Mechanics 2006, Svratka, Institute of Theoretical and Applied Mechanics AS CR, 2006, CD-ROM. (in Czech)

[18] Volek, J., Soukup, J., Šimsová, J., About One Solution of Some Algebraic Equations of Order $n = 2p$, So-called Frequency Equations for the Vibration of Mechanical Systems of Two and Many Degrees of Freedom p, Proceedings of the 2nd International Conference Dynamics of Rigid and Deformable Bodies 2004, Ústí nad Labem, Department of technology and production management, University of J. E. Purkyně in Ústí nad Labem, 2004, pp. 89–105. (in Czech)

14

Dynamic wheelset drive load of the railway vehicle caused by short-circuit motor moment

V. Zeman[a,*], Z. Hlaváč[a]

[a]*Faculty of Applied Sciences, University of West Bohemia, Univerzitní 22, 306 14 Plzeň, Czech Republic*

Abstract

The paper deals with mathematical modelling of dynamic response of the railway vehicle wheelset drives caused by short-circuit traction motor torque. The individual wheelset drive with hollow graduated shaft is one of subsystems of the two-axled vehicle bogie with two wheelset drives. The model respects the viscoelastic suspension of the both engine stators with gear housings mounted on the bogie frame and all the other couplings among bogie drive components. The dynamic response is investigated in dependence on longitudinal creepage and forward velocity of the vehicle at the moment of the sudden short-circuit in one asynchronous traction motor. The method is applied to bogie of the electric locomotive developed for speed about 200 km/h by the company ŠKODA TRANS-PORTATION, s. r. o. The wheelset drive vibration is confronted with stability conditions of the whole bogie.

Keywords: railway vehicle bogie, short-circuit moment, stability conditions, dynamic load

1. Introduction

Dynamic properties of individual wheelset drives of railway vehicles are usually investigated using torsional models, as it was shown e.g. in [9, 11]. These models, however, do not enable investigation of dynamic load of wheelset drive components (Fig. 1) affected by spatial vibrations of traction motor (TM), gear housing with gears (G), hollow shaft (H) embrasing the wheelset axle and wheelset (W). The spatial vibrations of bogie components affect shaft torques, forces transmitted by gearing, clutches, viscoelastic supports between traction motors with gear housings and the bogie frame (BF) and creep forces acting at the contact patches between rails and wheels. Hence, complex models of railway vehicles or their components, presented e.g. in books [7, 13], in the latest works [6, 10] and there cited papers, were developed. The complex model of the railway vehicle bogie (Fig. 2) with radial, lateral, torsional and bending elastic wheels (Fig. 3) was developed by authors [8] for the purpose of optimization of design parameters in term of dynamic response caused by irregularities of the track geometry and by the polygonalized running surface of the wheels. None of complex and detailed models of the railway vehicle bogie has been used for determination of a dynamic response caused by short-circuit moment in one traction motor.

In this paper the detailed linearized model of the two-axled bogie with two individual wheelset drives is used for investigation of this extreme phenomenon in dependence on operational conditions at the short-circuit instant in one traction motor.

---

*Corresponding author. e-mail: zemanv@kme.zcu.cz.

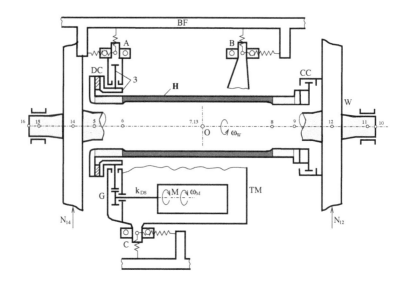

Fig. 1. Scheme of wheelset drive with a hollow shaft

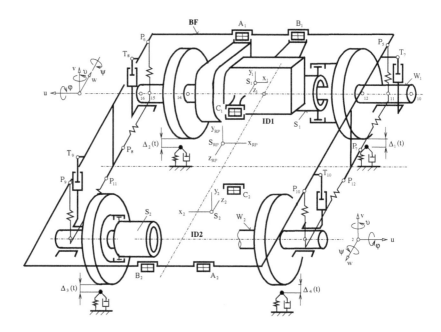

Fig. 2. Scheme of the bogie

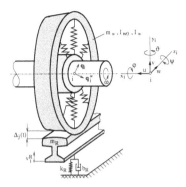

Fig. 3. Scheme of the elastic wheel

2. Mathematical model of the bogie

The development of the mathematical model of the bogie with rigid wheels was presented in the paper [15] and detailed in the research report [14]. The mathematical model of the bogie with elastic wheels was derived in configuration space [8]

$$q(t) = [q_{ID1}^T(t), q_{BFCB}^T(t), q_{ID2}^T(t)]^T \tag{1}$$

of dimension 189, where subvectors correspond to three subsystems – two individual wheelset drives and bogie frame linked by the secondary suspension to a half of the car body. Each individual drive (subscripts ID1 and ID2) is composed from rigid mass components (see Fig. 2) – rotor of traction motor, driving and driver gear, stator of traction motor with gear housing. These components are coupled by massless viscoelastic couplings – driving shaft with torsional stiffness k_{DS}, gearing with mesh stiffness k_G, disc clutch (DC) characterized by diagonal stiffness matrix (stiffnesses with one subscript are translational and with double subscript are flexural)

$$K_{DC} = \mathrm{diag}[k_x, k_y, k_z, k_{xx}, k_{yy}, k_{zz}],$$

rubber silent blocks with centres of elasticity A_1, B_1, C_1 (for ID1) and A_2, B_2, C_2 (for ID2) characterized by translation stiffnesses arranged in diagonal matrix $K_{SB} = \mathrm{diag}[k_x, k_y, k_z]$, disc clutch (DC), claw clutch (CC) with torsional stiffnesses k_{DC} and k_{CC} and railway balast (rail, railpad, sleeper and balast) reduced to a single mass-spring-damper system [5] defined by mass, stiffness and damping parameters m_R, k_R, b_R. The composite hollow shafts and wheelset axes are considered to be one-dimensional continua and are discretized by FEM. Their node displacements are expressed by the vector (see Fig. 2)

$$q_i = [u_i, v_i, w_i, \varphi_i, \vartheta_i, \psi_i]^T, \; i = 5, \ldots, 16. \tag{2}$$

The rigid discs of clutches, journals and wheels are mounted at nodes $i = 5$ (DC), $i = 9$ (CC), $i = 11, 15$ (journals) and $i = 12, 14$ (wheels). The flexible connection between the wheel rims and the wheel discs (Fig. 3) can be represented by massless springs and dampers [2]. Each wheel rim may undergo lateral, vertical, longitudinal, torsional, yaw and roll motion described by the displacement vector

$$q_i^w = [u_i^w, v_i^w, w_i^w, \varphi_i^w, \vartheta_i^w, \psi_i^w]^T, \; i = 12, 14 \tag{3}$$

for both individual drives ID1 and ID2. Individual drives are placed centrally symmetrical in the bogie. The rigid bogie frame is linked by secondary suspension P_1–P_4 and dampers T_1–T_6 with a half of car body (Fig. 4).

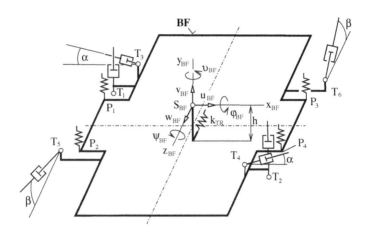

Fig. 4. Model of the bogie frame with the secondary suspension

The mathematical model of the bogie has the form [8]

$$M\ddot{q}(t) + B\dot{q}(t) + Kq = f_G + f_M(\dot{q}, t) + f_{R,W}(q, \dot{q}, t), \qquad (4)$$

where matrices have the block-diagonal structure

$$
\begin{aligned}
M &= \operatorname{diag}[M_{ID}, M_{BFCB}, M_{ID}], \\
B &= \operatorname{diag}[B_{ID}, B_{BFCB}, B_{ID}] + B_{D,BF} + B_{W,BF}, \\
K &= \operatorname{diag}[K_{ID}, K_{BFCB}, K_{ID}] + K_{D,BF} + K_{W,BF}
\end{aligned}
\qquad (5)
$$

corresponding to subsystems. Matrices $B_{D,BF}$ and $K_{D,BF}$ describe the viscoelastic supports of the stators with gear housings of both traction motors to the bogie frame in silent blocks. Matrices $B_{W,BF}$ and $K_{W,BF}$ describe damping and stiffnesses of the primary suspension at points T_7 to T_{10} (damping) and P_5, P_6, P_9, P_{10} (stiffness) and the longitudinal wheelset guide between journal boxes and the bogie frame at points P_7, P_8, P_{11}, P_{12} (see Fig. 2). The vector f_G expresses all gravitational forces and the vector $f_M(\dot{q}, t)$ expresses the motor driving torques. The vector $f_{R,W}(q, \dot{q}, t)$ includes contact forces between rails and wheel rims affected by track or wheel surface deviations $\Delta_j(t)$, $j = 1, 2, 3, 4$ (see Fig. 2 and Fig. 3).

3. Linearized mathematical model of the bogie

To analyze the dynamic response of the bogie caused by the sudden short-circuit for instance in traction motor ID1 we neglect track and wheel irregularities $\Delta_j(t) = 0$, $j = 1, 2, 3, 4$. The torque characteristics of the fellow asynchronous traction motor of ID2 is linearized in the neighbourhood of the state before short-circuit

$$M_{ID2} = M(s_0, v) - b_M \Delta\dot{\varphi}_1^{(ID2)}, \qquad (6)$$

where b_M is the slope of the traction motor characteristics and $\Delta\dot{\varphi}_1^{(ID2)}$ is disturbance angular velocity of the rotor with respect to rotation corresponding to vehicle forward velocity v and longitudinal creepage s_0 of all wheels. The motor torque of both electric motors in a state of static equilibrium is

$$M(s_0, v) = 2\mu(s_0, v)N_0 r_0/p, \qquad (7)$$

where $\mu(s_0, v)$ is longitudinal creep coefficient [11], N_0 is static vertical wheel force, r_0 is wheel radius in central position and $p = \frac{\omega_M}{\omega_W}$ is speed ratio.

Longitudinal $T_{i\,ad}$, lateral $A_{i\,ad}$ creep forces and spin torque $M_{i\,ad}$ acting at the contact patches between rails and wheels can be expressed as

$$
\begin{aligned}
T_{i\,ad} &= \mu(s_i, v)N_i\,, & (8)\\
A_{i\,ad} &= b_{22}(\dot{u}_i^w + r_i\dot{\psi}_i^w) + b_{23}\dot{\vartheta}_i^w\,, & (9)\\
M_{i\,ad} &= -b_{23}(\dot{u}_i^w + r_i\dot{\psi}_i^w) + b_{33}\dot{\vartheta}_i^w\,, i = 12, 14 & (10)
\end{aligned}
$$

for ID1 and ID2. The longitudinal creep force is expressed in dependence on actual longitudinal creep coefficient $\mu(s_i, v)$ and on vertical wheel force

$$
N_i = N_0 - (m_R\ddot{v}_i^w + b_R\dot{v}_i^w + k_R v_i^w)\,. \tag{11}
$$

The longitudinal creep coefficient $\mu(s_i, v)$ depends on longitudinal creepage defined by

$$
s_i = s_0 + \frac{\pm\dot{w}_i^w \mp r_i\Delta\dot{\varphi}_i^w}{v}\,, \quad s_0 = \frac{r_0\omega_W}{v}\,, \quad i = 12, 14\,. \tag{12}
$$

whereas upper signs correspond to wheelset W_1 and lower signs to wheelset W_2, which rotate with angular velocity ω_W before the sudden shot-circuit. The vertical wheel forces are expressed in dependence on vertical displacements v_i^w, velocities $\dot{v}_i^w$ and accelerations $\ddot{v}_i^w$ of wheel rim mass centre. The whole track structure (rail, railpad, sleepe and balast) is reduced to a single mass-spring-damper system [5] defined by mass, stiffness and damping parameters m_R, b_R, k_R figuring in term for N_i. The lateral creep force and the spin torque about vertical axis depend on linearized creep coefficients b_{ij}, actual wheel radius r_i and wheel rim mass centre velocities, marked with subscript w (see Fig. 3). The creep coefficients were calculated using Kalker's theory [7] for static vertical wheel force N_0.

To analyze the modal properties, stability conditions and vibration of the bogie, the longitudinal creep characteristics defined in [3, 12] and presented in Fig. 5 are linearized in the neighbourhood of a state before short-circuit in the form

$$
\mu(s_i, v) = \mu_0(s_0, v) + \left[\frac{\partial\mu}{\partial s_i}\right]_{s_i=s_0}(s_i - s_0)\,. \tag{13}
$$

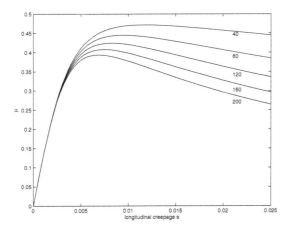

Fig. 5. Creep characteristics

The linearized longitudinal creep forces can be then expressed for $N_i = N_0$ and $r_i = r_0$ as

$$T_{i\,ad} = \mu(s_o, v)N_0 + b_{11}(\pm \dot{w}_i^w \mp r_0 \Delta \dot{\varphi}_i^w)\,, \; i = 12, 14\,, \tag{14}$$

where

$$b_{11} = \frac{N_0}{v}\left[\frac{\partial \mu}{\partial s_i}\right]_{s_i = s_0}. \tag{15}$$

If the static equilibrium is disturbed by short-circuit moment $M_C(t)$ of the traction motor in ID1, the vector of generalized coordinates is expressed as a sum of static and dynamic displacements

$$q(t) = q_0 + \Delta q(t)\,. \tag{16}$$

After expressing of the motor torque in ID2 according (6) and the creep forces according (8) to (15), the force vectors on right side of equation (4) can be written as

$$f_G + f_M(\dot{q}, t) + f_{R,W}(q, \dot{q}, t) = f_0 - [B_M + B_{ad}(s_0, v)]\Delta \dot{q}(t) + \Delta f(t)\,. \tag{17}$$

The vector $f_0 = K q_0$ expresses static force effects before the sudden short-circuit. The diagonal matrix B_M has nonzero elements b_M on positions corresponding to rotor and stator angular velocities of the traction motor of ID2 in vector $q(t)$. The block diagonal matrix of all creep forces

$$B_{ad}(s_0, v) = \mathrm{diag}[\ldots, \bar{B}_{ad} \ldots, \bar{B}_{ad} \ldots, \bar{B}_{ad} \ldots, \bar{B}_{ad}] \tag{18}$$

has nonsymmetrical blocks

$$\bar{B}_{ad} = \begin{bmatrix} b_{22} & 0 & 0 & 0 & b_{23} & r_0 b_{22} \\ 0 & 0 & 0 & 0 & 0 & 0 \\ 0 & 0 & b_{11} & -r_0 b_{11} & 0 & 0 \\ 0 & 0 & -r_0 b_{11} & r_0^2 b_{11} & 0 & 0 \\ -b_{23} & 0 & 0 & 0 & b_{33} & -r_0 b_{23} \\ r_0 b_{22} & 0 & 0 & 0 & r_0 b_{23} & r_0^2 b_{22} \end{bmatrix}, \tag{19}$$

which are localized on positions corresponding to wheel rim displacement vectors q_i^w, $(i = 12, 14$ for both wheelsets) in vector of generalized coordinates $q(t)$. The linearized mathematical model of the bogie according (4), (16) and (17) can be written in perturbance coordinates in the neighbourhood of the static equilibrium as

$$M\Delta\ddot{q}(t) + [B + B_M + B_{ad}(s_0, v)]\Delta\dot{q}(t) + K\Delta q(t) = \Delta f(t)\,. \tag{20}$$

The excitation (perturbation) vector $\Delta f(t)$ has nonzero components $M_C(t)$ on positions corresponding to angular displacements of the rotor and stator of the traction motor in ID1 in vector of generalized coordinates $q(t)$.

4. Stability conditions of the bogie

The stability conditions of the bogie with rigid wheels were investigated in [15]. In this paper, for the purpose of association with dynamic response caused by short-circuit traction motor, the complex linearized autonomous mathematical model (20) can be used for stability analysis. Eigenvalues are defined by eigenvalue problem solution

$$[\lambda_\nu N(s_0, v) + P]u_\nu = 0 \tag{21}$$

in the state space $u = [\Delta \dot{q}^T,\ \Delta q^T]^T$, defined by matrices

$$N(s_0, v) = \begin{bmatrix} 0 & M \\ M & B + B_M + B_{ad}(s_0, v) \end{bmatrix}, \quad P = \begin{bmatrix} -M & 0 \\ 0 & K \end{bmatrix}. \qquad (22)$$

The eigenvalues λ_ν depend on operational parameters s_0, v, N_0 and on the slope b_M of torque characteristic of the traction motor in ID2 at the instant of the short-circuit in the motor of ID1. In the first step we perform eigenvalue problem solution for different longitudinal creepage $s_0 = 0.005$ (stable state) and $s_0 = 0.014$ (unstable state) corresponding to motor torque $M_0(s_0, v) = 12\,800$ [Nm] for forward vehicle velocity $v = 120$ [km/h] and $N_0 = 10^5$ [N].

Table 1. Eigenvalues of bogie

eigenvalues	s_0=0.005, v=120 km/h		s_0=0.014, v=120 km/h	
sequence	complex	real	complex	real
1	$-0.032 \pm i1.262$	$-1.11 \cdot 10^{-11}$	$-0.029 \pm i1.262$	**7.031**
2	$-0.041 \pm i2.694$	$-2.73 \cdot 10^{-11}$	$-0.060 \pm i2.689$	**0.062**
3	$-0.199 \pm i5.425$	$-0.015\,6$	**0.076** $\pm i5.374$	$-1.17 \cdot 10^{-12}$
4	$-0.056 \pm i5.449$	$-0.027\,1$	$-0.031\,8 \pm i5.443$	$-4.36 \cdot 10^{-9}$
5	$-1.583 \pm i6.176$	$-0.071\,7$	$-1.583 \pm i6.174$	$-0.015\,8$
6	$-0.164 \pm i8.503$	-1.611	$-0.164 \pm i8.505$	$-0.029\,9$
7	$-0.289 \pm i8.583$	-1.790	**0.416** $\pm i9.126$	$-0.071\,7$
8	$-3.282 \pm i9.183$	-5.147	**1.111** $\pm i9.535$	-1.612
9	$-3.958 \pm i12.208$	-6.221	**0.911** $\pm i12.140$	-5.438
10	$-4.010 \pm i12.58$	-14.521	**0.809** $\pm i12.162$	-6.254

Table 2. Characteristics of the vibration mode shapes

ν	Imλ_ν[Hz]	Dominant vibrations of bogie components accordant with mode shapes
1	1.262	vertical of CB in phase with BF and TMs of both IDs
2	2.694	torsion of TM rotor of ID1 with gear transmission, twisting of disc clutch
3	5.425	torsion of both gears and BF pitch, gearing deformations
4	5.449	vertical of BF with both TMs in phase
5	6.176	lateral and roll of BF with both TMs in phase
6	8.503	lateral of both TMs in opposite phase
7	8.584	combined longitudinal-pitch of W2 and combined BF roll-pitch
8	9.183	combined longitudinal-pitch of W1 and combined BF roll-pitch
9	12.208	combined longitudinal-yaw of W2
10	12.580	combined longitudinal-yaw of W1
58	298.4	torsion of ID1 pinion gear, twisting of ID1 driving shaft

BF... bogie frame
CB... car body
TM... traction motor
W1... wheelset of ID1
W2... wheelset of ID2

These values describe the possible operational state of the particular electric locomotive at the instant of the sudden short-circuit [4]. The first ten pairs of complex conjugate eigenvalues sequenced according to magnitude of imaginary parts and ten real eigenvalues sequenced from smallest values is presented in Table 1. Vibration mode shapes, corresponding to complex conjugate eigenvalues for $s_0 = 0.005$, are characterized in Table 2 in agreement with dominant vibrations and deformations of bogie components. Aperiodic mode shapes, corresponding to negative real eigenvalues , have no importance for dynamic load. Positive real parts of eigenvalues reflect system instability. This instability corresponding to positive real part of the complex conjugate eigenvalues is flutter type (see five eigenvalues for $s_0 = 0.014$ in Table 2) and instability of a divergence type corresponds to real eigenvalues (see the first and second real eigenvalues for $s_0 = 0.014$). Obviously the system is stable for longitudinal creepage $s_0 = 0.005$ and unstable for large creepage $s_0 = 0.014$.

As an illustration, we present the dependence of real and imaginary parts of eight lowest eigenvalues on the longitudinal creepage s_0 for the vehicle velocity $v = 120$ km/h in Fig. 6. The stability limit for $v = 120$ km/h is defined by creepage $s_0 = 0.008\,2$. We take a note that for higher vehicle velocity the limiting creepage is smaller (for $v = 200$ km/h $s_0 = 0.006\,9$).

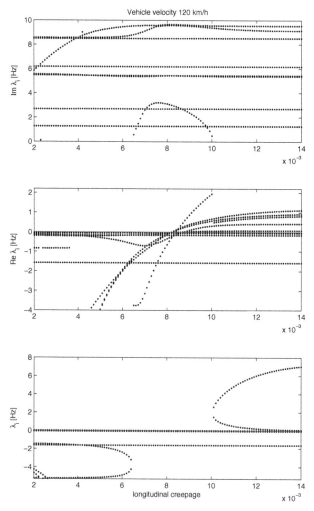

Fig. 6. Dependence of the imaginary parts (top), real parts (mid) of complex conjugate eigenvalues and real eigenvalues (bottom) on longitudinal creepage s_0

5. Dynamic response caused by short-circuit motor torque

The short-circuit motor torque in the air-space of the particular traction motor was calculated in ŠKODA ELECTRIC, a. s., in dependence on time [4]. This dependence can be well approximated in perturbation coordinates of the model (20) by function (Fig. 7)

$$M_C(t) = -M(s_0, v)H(t) - M_0 e^{-D\omega t} \sin[\omega(t - \Delta t)], \tag{23}$$

where $M(s_0, v)$ is the traction motor torque in a state of the static equilibrium just before short-circuit, $H(t)$ is Heaviside function and the oscillating short-circuit torque is defined by amplitude M_0, frequency ω, shift phase $\omega \Delta t$ and short-circuit torque decay $D\omega$. The total motor torque after short time (here 0.2 [s]) is equal zero (in perturbation coordinates is $-M(s_0, v)$).

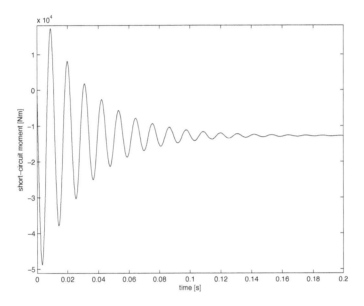

Fig. 7. Function approximating the disturbance by short-circuit moment in ID1

As an illustration the time behaviour in interval $t \in \langle 0; 2 \rangle$ [s] of the dynamic torques and forces transmitted by chosen linkages of the wheelset drive of ID1 for operational parameters $s_0 = 0.005$ and $v = 120$ km/h at the instant of the short-circuit are presented in Fig. 8 to Fig. 10.

The frequency $f = \omega/2\pi = 90$ [Hz] of the oscilating short-circuit torque and eigenfrequencies $f_2 = 2.69$ [Hz] and $f_{58} = 298.4$ [Hz], corresponding to mode shapes (see Table 2) characterized by torsion deformation of the flexible disc clutch (f_2) and driving shaft (f_{58}), show up as dominant. The identical values of the wheelset drive of ID2 are multiple smaller. The short-circuit motor torque causes an extreme load of the driving shaft torque M_{DS} approximately 50 % of its maximal value in time 0.18 [s] (Fig. 8). Maximal disc clutch torque $M_{x\,DC}$ in the same time is cca 10^5 [Nm] (Fig. 9) which means greater value than the maximal adhesion wheelset moment $M_{ad} = 2N_0 r \mu_{\max}(v) = 0.52 \cdot 10^5$ [Nm] for $v = 120$ [km/h].

Fundamentally worseness arises in the event of the short-circuit at large longitudinal creepage in the downward section of the creep characteristic (see Fig. 5). As an illustration the time behaviour of the driving shaft torque M_{DS} and of the disc clutch torque is shown in Fig. 11 for operational parameters $s_0 = 0.014$ and $v = 120$ [km/h] at the moment of the short-circuit. This model example illustrates relationship between modal properties and dynamic response of

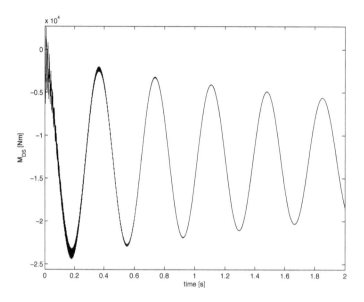

Fig. 8. Driving shaft torque

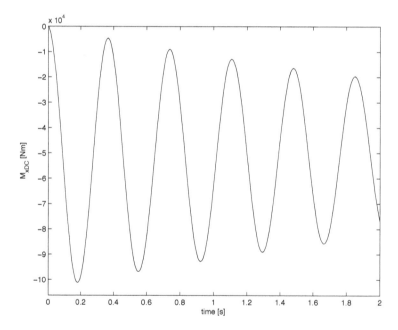

Fig. 9. Disc clutch moment

the unstable system and at once gives account of failure cause experimentally evidenced during testing operation of the real railway vehicle. Such large longitudinal creepage occurs e.g. in the case of a wet or a contaminated face of the rail. In consequence of system unstability (real parts of five complex conjugate eigenvalues and two real values are positive – see Table 1) the both above-mentioned torques continuously increse until the antislip protection equipment is activated. The activation time should not be greater than cca 0.35 [s].

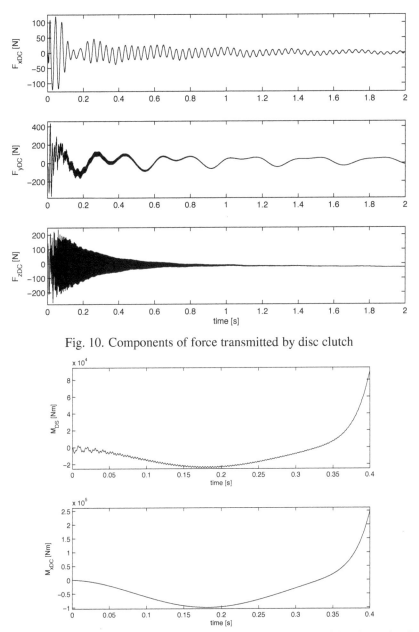

Fig. 10. Components of force transmitted by disc clutch

Fig. 11. Driving shaft torque (top) and disc clutch torque (bottom) caused by short-circuit moment in ID1 in unstable state ($s_0 = 0.014$; $v = 120$ km/h)

6. Conclusion

The paper presents the original mathematical modelling method and computer simulation of dynamic load of the wheelset drive caused by short-circuit motor moment. The detailed complex model of the railway vehicle bogie with two individual wheelset drives was used for studying of this extreme phenomenon. The model respects spatial vibrations of the traction motors, gear housings, hollow shafts, wheelsets, bogie frame and viscoelastic coupling among bogie components and among wheel rims and wheel discs, respectively. The mass, stiffness and damping of the rail ballast and the longitudinal, lateral and spin linearized creep forces in wheel-rail contacts are respected.

The dynamic response of the wheelset drive depends strongly on longitudinal creepage of wheels while on the forward locomotive velocity at the moment of the short-circuit has insignificant influence. In the event of the short-circuit in one traction motor at large longitudinal creepage the dynamic load of this wheelset drive extremely increases until activation of the antislip protection. The dynamic load of the normally working wheelset drive is low.

The developed software in MATLAB code enables graphically record time behaviour of the arbitrary generalized coordinates and with small software modification also the arbitrary forces transmitted by linkages between bogie components.

Acknowledgements

This paper includes partial results from the Research Project MŠMT 1M0519 – Research Centre of Rail Vehicles supported by the Czech Ministry of Education, Youth and Sports.

References

[1] Byrtus, M., Zeman, V., Hlaváč, Z., Modelling and modal properties of an individual wheelset drive of a railway vehicle, Sborník XVIII. mezinárodné konferencie Súčasné problémy v kolajových vozidlách – PRORAIL2007, Žilina.

[2] Claus, H., Schiehlen, W., System Dynamics of Railcars with Radial and Laterelastic Wheels, in System Dynamics and Long Term Behaviour of Railway Vehicles, Track and Subgrade (K. Popp and W. Schiehlen eds.), Springer, 2003, p. 65–84.

[3] Čáp, J., Some aspects of uniform comment of adhesion mechanism, Proceedings of VŠCHT Pardubice works, 1993, p. 26–35 (in Czech).

[4] Dvořák, P., Calculated values of the short-circuit moment in the air-space of the traction motor ML 4550 K/6, Documentation of ŠKODA ELECTRIC, a.s., Plzeň, 2009.

[5] Feldman, V., Kreuzer, E., Pinto, F., Schlegel, V., Monitoring the dynamic of railway track by means of the Karhunen-Loeve-transfomation, in System Dynamics and Long Term Behaviour of Railway Vehicles, Track and Subgrade (K. Popp and W. Schiehlen eds.), Springer, 2003, p. 231–246.

[6] Fleiss, R., Laufverhalten des quergefederten Tatzlagerantriebs, ZEVrail Glasers Annalen 130 (6,7) (2006) 264–269.

[7] Grag, V., Dukkipati, R., Dynamics of railway vehicle systems, London Academic Press, 1984.

[8] Hlaváč, Z., Zeman, V., Optimization of the railway vehicle bogie in term of dynamic, Applied and Computational Mechanics, 1 (3) (2009) 39–50.

[9] Jöckel, A., Aktive Dämpfung von Ratterschwingungen in Antriebstrang von Lokomotive mit Drehstrom-Antriebstechnik, ZEV + DET Glas. Ann., 125 (5) (2001) 191–204.

[10] Knothe, K., Systemdynamik und Langzeitverhalten von Shienenfahrzeugen, Gleis und Untergrund, ZEVrail Glasers Annalen 131 (4) (2007) 154–168.

[11] Lata, M., Modelling of transient processes in torsion drive system of railway vehicle, Scientific Papers of the University of Pardubice, Series B, 2003, p. 45–58 (in Czech).

[12] Lata, M., Dynamic processes in electric locomotive drive on rise of the wheelset slip, Book of Extended Abstracts of the Conference Engineering Mechanics 2004, Svratka, p. 165–166, (full paper on CD-rom).

[13] Popp, K., Schiehlen, W. (Eds.), System Dynamics and Long-Term Behaviour of Railway Vehicles, Track and Subgrade, Springer-Verlag, 2003.

[14] Zeman, V., Hlaváč, Z., Byrtus, M., Modelling of wheelset drive vibration of locomotive 109E, Research report n. H2,H6-06-01/2006, Research Centre of Rail Vehicles, Plzeň, 2007 (in Czech).

[15] Zeman, V., Hlaváč, Z., Byrtus, M., Modellig and modal properties of the railway vehicle bogie with two individual wheelset drives, Applied and Computational Mechanics, 1 (1) (2007) 371–380.

15

Implementation of skeletal muscle model with advanced activation control

H. Kocková[a,*], R. Cimrman[b]

[a] *New Technologies – Research Centre in the Westbohemian Region, University of West Bohemia, Univerzitní 22, 306 14 Plzeň, Czech Republic*

[b] *Department of Mechanics, Faculty of Applied Sciences, University of West Bohemia, Univerzitní 22, 306 14 Plzeň, Czech Republic*

Abstract

The paper summarizes main principles of an advanced skeletal muscle model. The proposed mathematical model is suitable for a 3D muscle representation. It respects the microstructure of the muscle which is represented by three basic components: active fibers, passive fibers and a matrix. For purposes of presented work the existing material models suitable for the matrix and passive fibers are used and a new active fiber model is proposed. The active fiber model is based on the sliding cross-bridge theory of contraction. This theory is often used in modeling of skeletal and cardiac muscle contractions. In this work, a certain simplification of the cross-bridge distribution function is proposed, so that the 3D computer implementation becomes feasible. The new active fiber model is implemented into our research finite element code. A simple 3D muscle bundle-like model is created and the implemented composite model (involving the matrix, passive and active fibers) is used to perform the isometric, concentric and excentric muscle contraction simulations.

Keywords: skeletal muscle, cross-bridge distribution, calcium activation, fiber architecture, composite model

1. Introduction

Computer simulations of various human body functions have already shown a wide applicability in many branches of medicine, sports or vehicle safety research. The human body modeling respects two concepts in general. The first one is based on a tissue microstructure and enables a detailed modeling of body parts, while the second one models a human body as a whole usually in the interaction with surroundings. These two concepts complement each other.

This paper deals with modeling of skeletal muscle tissue and focuses on the detailed description of active properties. It provides a compact summary of the investigated problem including the relations presented in [11] completed by several essential conditions. The model is implemented into the MAFEST (Matlab Finite Element Simulator) software that is being developed at our laboratory, see [2]. MAFEST is a modular software suitable for various modeling tasks in biomechanics. Recently it has been used for implementation of smooth and cardiac muscle tissue modeling, together with the related sensitivity algorithm for material parameters identification, cf. [3, 17, 18]. The identification algorithm was used for smooth muscle tissue [13] and kidney [4] parameter determination.

The active fiber model is based on the cross-bridge kinetics (sometimes also called theory of sliding filaments) introduced firstly by Huxley [9] which was further extended or used by

---

*Corresponding author. e-mail: hcechov@ntc.zcu.cz.

many other authors, cf. [6, 15, 19]. The paper introduces an advanced active fiber model which in connection with a proper model of passive fibers and matrix gives a suitable 3D composite skeletal muscle representation.

2. Anatomical background

Following paragraphs give a brief summary of a skeletal muscle anatomy and a physiology of muscle contraction ease to a reader the orientation in a number of terms.

2.1. Skeletal muscle structure

Skeletal muscles can be described from the macroscopic and the microscopic points of view. We start with the macroscopic one. A long multinucleate fiber is a basic constructive element of each skeletal muscle that is important for this approach. Muscle fibers are then joined to a primary bundle and the parallel oriented bundles form a muscle.

At the microscopic level the fiber can be further divided into *myofibrils*, each containing a number of *filaments*. The cytoplasm, often called *sarcoplasm*, of the muscle filamentum involves many actual contractile elements called *sarcomeres*. The basis of each sarcomere is created by the thin filaments containing the protein *actin* and the thick filaments containing the protein *myosin*. They fit together and create so called *cross-bridges* which enable a mutual movement causing the contraction of the sarcomere, see fig. 1, left. The thin filaments of adjacent sarcomeres are anchored in so called Z-discs. The regular structure of Z-discs is responsible for a striated appearance of the skeletal muscle. The sarcomere structure is outlined in fig. 1, right.

2.2. Contraction cycle

The contraction occurss when the cross-bridges between actin and myosin bind and generate a force causing the thin filaments to slide along the thick filaments. According to the sliding filament theory the sarcomere force depends on the amount of overlaps of the actin and myosin filaments.

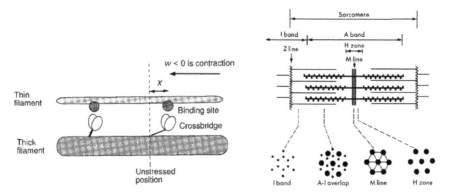

Fig. 1. Bonding of a cross-bridge (left), sarcomere structure(right); taken from [10]

The optimum sarcomere (fiber, muscle) length is the length when the overlap of the actin and myosin filaments enables the interaction of all cross-bridges and consequently the generated force touches its peak. With increasing muscle length the number of cross-bridges decreases. For lengths smaller than the optimum length, actin filaments overlap and the number of cooperating cross-bridges again decreases.

The passive force contributes also to the total force for lengths over a passive slack length, i.e. the smallest length when any force is exerted under passive conditions. The contraction is a very complicated cycle, which stands upon the creation and conversion of ATP and on the transfer of calcium ions, cf. [1, 10].

3. Composite model of skeletal muscle

The proposed model consists of a matrix, active fibers and passive fibers. A new active fiber model, whose incomplete version was introduced in [11], is implemented. The hyperelastic material model for the matrix and viscoelastic passive fibers with linear elastic response are chosen from the MAFEST library. The proposed active fiber model arises from a sliding cross-bridge theory of contraction, cf. [10], which is often used in skeletal and cardiac muscle modeling.

3.1. General setting

Below we use the following notation: vectors: $\square$ with components $\{\square_i\}$, $i = 1, 2, 3$; second order tensors: $\underline{\square}$ with components $\{\square_{ij}\}$, $i, j = 1, 2, 3$.

The composite model assumes that at any point in the material the properties of the solid reflect the microstructure. The microstructure of muscle is characterized by three basic components: *active fibers* representing bundles of muscle cells, *passive fibers* corresponding to collagen and elastin fibers and finally a *matrix* substituting the amorphous extracellular substance. These components are supposed to occupy an infinitesimal volume according to the volume fractions denoted by ϕ^a, ϕ^p and ϕ^m respectively, so that

$$\phi^a + \phi^p + \phi^m = 1. \tag{1}$$

All the components contribute then to the total stress in proportion to their volume fractions, cf. [16]. Large deformations exhibited by the muscle tissue are described using the total Lagrangian formulation (TLF). The total second Piola–Kirchhoff stress tensor $\underline{S}$ is represented by the summation of the contributions of each component which are proportional to the associated volume fractions:

$$\underline{S} = \phi^m \underline{S}^m + \phi^p \underline{\tau}^p + \phi^a \underline{\tau}^a, \tag{2}$$

where $\underline{S}^m$ is the matrix stress, $\underline{\tau}^p$ and $\underline{\tau}^a$ are the stress tensors of fibrous components. Similar composite model types are described for example in [8]. At any point of the continuum the model enables both active and passive fibers to be distributed in several preferential directions k in which the tension can be transmitted, k belongs to the index set I^a in case of the active fibers and to the index set I^p for the passive fibers. The kth direction is defined by the unit vector $v^k = (v_1^k, v_2^k, v_3^k)$ related to the undeformed configuration. A quantity of active fibers in the kth direction is proportional to the volume fraction ϕ_a^k, $\sum_{k \in I^a} \phi_a^k = 1$ and analogously for the passive fibers. Using the directional tensor $\omega_{ij}^k = v_i^k v_j^k$ the stress tension of fibrous components can be expressed as

$$\underline{\tau}^p = \sum_{k \in I^p} \phi_p^k \tau_p^k \underline{\omega}^k, \qquad \underline{\tau}^a = \sum_{k \in I^a} \phi_a^k \tau_a^k \underline{\omega}^k \tag{3}$$

(indexes ij omitted). Hence the fibrous components introduce *strong anisotropy* to the model.

We are interested mainly in quasi-static solutions and we omit in the following the inertia terms. Assuming the solution up to the step n to be known, we seek the solution at the step $n + 1$ satisfying the equilibrium equation

$$\int_{\Omega^{(0)}} \underline{S}^{(n+1)} : \delta \underline{E}(\underline{v}) \, d\Omega - L^{(n+1)}(\underline{v}) = 0, \quad \forall \underline{v} \in V^3, \tag{4}$$

where $\underline{E}$ is the Green deformation tensor, L represents linear loads and V is a suitable function space. The Green deformation tensor can be expressed using the deformation gradient $\underline{\underline{F}}$ as $\underline{\underline{E}} = 1/2(\underline{\underline{F}}^T\underline{\underline{F}} - \underline{\underline{I}}) = 1/2(\underline{\underline{C}} - \underline{\underline{I}})$. Here $\underline{\underline{F}}$ relates the spatial coordinates $\underline{x}$ to the material ones $\underline{X}$ by $\underline{\underline{F}} = \frac{\partial \underline{x}}{\partial \underline{X}}$, $\underline{\underline{C}}$ is the right Cauchy-Green deformation tensor, $\underline{\underline{C}} = \underline{\underline{F}}^T\underline{\underline{F}}$ and $\underline{\underline{I}}$ is the identity.

3.2. Matrix

A rather simple hyperelastic material model is chosen for the matrix, substituting the extracellular substance, since only small contribution of the matrix to the mechanical behavior of the complex tissue is considered, which was proved sufficient in similar models, cf. [5] and [14]. To respect the nearly incompressibility of soft tissues it is convenient to split the $\underline{\underline{S}}^m$ into the effective (shear) part $\underline{\underline{S}}^{eff}$ and the volumetric (pressure) part $-pJ\underline{\underline{C}}^{-1}$ with $J = \det(\underline{\underline{F}})$. The incompressibility is treated using common definition $p = -K(J - 1)$, where K is the bulk modulus.

In our model the matrix is represented by the hyperelastic material, the second Piola Kirchhof stress tensor can be obtained as

$$\underline{\underline{S}}^m = \frac{\partial W}{\partial \underline{\underline{E}}}. \tag{5}$$

W denotes the strain energy which can have various forms according to a type of material. The simplest hyperelastic material, the neo-Hookean, was chosen:

$$\underline{\underline{S}}^m = \underline{\underline{S}}^{eff} - pJ\underline{\underline{C}}^{-1} = \mu J^{-\frac{2}{3}}(\underline{\underline{I}} - \frac{1}{3}\mathrm{tr}(\underline{\underline{C}})\underline{\underline{C}}^{-1}) + K(J - 1)J\underline{\underline{C}}^{-1}, \tag{6}$$

where the shear modulus μ and the bulk modulus K are the material parameters.

3.3. Passive fibers

The passive fibers characterize the collagen and elastin network in the muscle tissue which shows nonlinear viscoelastic behavior. The fibers transmit only tension. The viscoelastic passive fibers with linear elastic response are chosen from the MAFEST material library [2]. This type of passive fibers is represented by the uniaxial three parametric Kelvin-Zener model. The viscoelastic stress τ^p (in the following the superscript p omitted for brevity) is a function of the elastic response σ and the internal stress-like variable q and it is expressed as

$$\tau = \sigma - q, \tag{7}$$

$$\dot{q} + \frac{1}{T_\epsilon}q = \frac{\gamma}{T_\epsilon}\sigma \tag{8}$$

where T_ϵ is the relaxation time and γ is the relaxation parameter. At the thermodynamic equilibrium ($\dot{q} = 0$) from 7 and 8 one can obtain an explicit formula for $\sigma(t)$

$$\sigma(t) = \frac{1}{1 - \gamma}\left(\tau(t) - \gamma\int_{t_0}^t e^{-\beta(t-s)}\frac{d\tau(s)}{ds}\,ds\right) \tag{9}$$

where $\beta = (1 - \gamma)T_\epsilon$. The elastic response is given always by the instantaneous strain $\varepsilon_p(t)$ obtained from displacement field. A very simple linear elastic response can be defined as

$$\sigma(t) = D\varepsilon_p(t)(t) \tag{10}$$

where D denotes a positive constant.

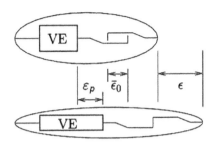

Fig. 2. Passive fiber

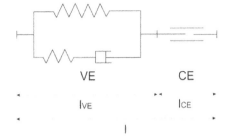

Fig. 3. Schema of the active fiber model

The collagen fibers are tangled in helical bundles in the undeformed tissue. Consequently they can transmit tension only after their straightening, while in compression they cannot transmit any load. This behavior is taken into account by the relative slack length parameter $\bar{\epsilon}_0$ depicted in Fig. 2 and expressed by the following relations:

$$\varepsilon_p(t) \geq \epsilon(t) - \bar{\epsilon}_0$$

$$\tau(t) \geq 0 \tag{11}$$

$$\tau(t) \cdot (\varepsilon_p(t) - \epsilon(t) + \bar{\epsilon}_0) = 0$$

where $\bar{\epsilon}_0 \geq 0$ allows taking into account the waviness of collagen fibers in released state and $\bar{\epsilon}_0 < 0$ corresponds to residual stresses.

3.4. Active fiber model

The active fiber model employs ideas from a sliding cross-bridge theory of contraction. This theory also known as the kinetic theory, was firstly proposed by A. F. Huxley already in 1957 and later it was extended by many other authors. The principles are based on modeling of bonding and debonding cross-bridges between actin and myosin filaments. Major activities were focused on the development of constitutive equations for description of the calcium activation and cross-bridge bonding/debonding process. Fundamental seems to be a work [19] where an approximation of a cross-bridge distribution function describing the kinetic theory was introduced.

The muscle tissue besides contains active filaments, as well the material embodying the viscoelastic behavior. Hence our model is constituted of the contractile element (CE) in series with viscoelastic element (VE), see fig. 3. The VE respects the Kelvin-Zener rheological model, the CE model arises from the Huxley type two state model.

3.5. Contractile part of active fiber

The constitutive equation of the contractile myofibers is based on the micro-mechanical Huxley model. The simplest model of Huxley type assumes that the cross-bridges have only two possible states: bonded and debonded. The bonded cross-bridge generates a force that causes shortening of the sarcomere. The distribution of attached cross-bridges with respect to their length ξ is given by the function $n(\xi, t)$:

$$\frac{\partial n(\xi, t)}{\partial t} - w(t)\frac{\partial n(\xi, t)}{\partial \xi} = r(t)f(\xi)\left(\alpha - n(\xi, t)\right) - g(\xi)n(\xi, t), \tag{12}$$

where w represents the macroscopic shortening velocity [20], α denotes the overlap function:

$$\alpha\left(\frac{\lambda_{CE}}{\lambda_{opt}}\right) = \begin{cases} 1 - 6.25\left(\frac{\lambda_{CE}}{\lambda_{opt}} - 1\right)^2 & \text{for } \frac{\lambda_{CE}}{\lambda_{opt}} \leq 1, \\ 1 - 1.25\left(\frac{\lambda_{CE}}{\lambda_{opt}} - 1\right) & \text{for } \frac{\lambda_{CE}}{\lambda_{opt}} > 1, \end{cases} \tag{13}$$

where λ_{CE} represents the ratio of instantaneous contractile element length l_{CE} and rest contractile element length L_{CE}, λ_{opt} is the optimal stretch. The function $f(\xi)$ is the attachment rate function and $g(\xi)$ is the detachment rate function of actin to myosin. Both of them are defined according to [19].

The activation factor r in (12) represents the fraction of sites on the actin filament that is activated. This activation occurs if the corresponding troponin site of the attachment sites is unlocked by two calcium ions. The activation factor r is determined from the chemical equilibrium between calcium and troponin and it is expressed as:

$$r = \frac{C^2}{C^2 + \mu C + \mu^2}, \tag{14}$$

with C the calcium concentration in the myofibrilar space normalized with respect to the maximum myofibrilar calcium concentration and μ the troponin-calcium reaction ratio constant. The rate of change of the normalized calcium concentration in the myofibrilar space is defined according to [7]:

$$\frac{\partial C}{\partial t} = \vartheta(c\nu - C), \tag{15}$$

where ϑ stands for the fiber-type dependent rate parameter, c is the calcium release parameter and ν ($\nu = f/f_{max}$) is the stimulation frequency f normalized by the tetanic stimulation frequency f_{max}.

In the Huxley theory a bound cross-bridge is supposed to behave like a spring. The contractile stress developed by all cross-bridges in a slice of half sarcomeres is:

$$\sigma_{CE} = K_A \int_{-\infty}^{\infty} \xi n(\xi, t)\,\mathrm{d}\xi = K_A Q_1(t) \tag{16}$$

with K_A a material constant. The term Q_1 is the first moment of the distribution function $n(\xi, t)$.

3.6. Simplified distribution-moment approximation

The equation (16) shows that the contractile stress does not depend on the exact shape of $n(\xi, t)$ but it is proportional to the first order distribution moment of $n(\xi, t)$. The kth distribution moment (DM) of $n(\xi, t)$ is generally defined as:

$$Q_k(t) = \int_{-\infty}^{\infty} \xi^k n(\xi, t)\,\mathrm{d}\xi \qquad k = 0, 1, 2, \ldots \tag{17}$$

This idea was first used by Zahalak and published in [19]. The DM method transforms the partial differential equation (12) to a set of first-order ordinary differential equations:

$$\frac{\partial Q_k}{\partial t} = \alpha r \beta_k - r\phi_{k1} - \phi_{k2} - kwQ_{k-1} \qquad k = 0, 1, 2, \ldots \tag{18}$$

with the integrals

$$\beta_k = \int_{-\infty}^{\infty} \xi^k f(\xi)\,d\xi, \quad \phi_{k1} = \int_{-\infty}^{\infty} \xi^k f(\xi)n(\xi,t)\,d\xi, \quad \phi_{k2} = \int_{-\infty}^{\infty} \xi^k g(\xi)n(\xi,t)\,d\xi. \quad (19)$$

If the approximation of $n(\xi,t)$ is chosen suitably then the system (18) becomes explicit for the moments Q_k.

In [19] the Gaussian approximation is chosen as a proper approximation of participating cross-bridge distribution. Thus $n(\xi,t)$ is characterized by its first three moments Q_0, Q_1, Q_2:

$$n(\xi,t) = \frac{Q_0}{\sqrt{2\pi}q(t)}e^{-\frac{(\xi-p)^2}{2q^2}}, \quad (20)$$

$$p = \frac{Q_1}{Q_0} \quad \text{and} \quad q = \sqrt{\frac{Q_2}{Q_0} - \left(\frac{Q_1}{Q_0}\right)^2}, \quad (21)$$

As a result a system of three first-order ordinary differential equations is obtained instead of partial differential equation.

Substituting the Gaussian approximantion of $n(\xi,t)$ into integrals ϕ_{k1} and ϕ_{k2} we obtain expressions involving the error function, see [19]. Their computer implementation is rather complicated. Hence it is convenient to further approximate the Gaussian approximation of $n(\xi,t)$ with a polynomial function. In [11] we proposed to approximate $n(\xi,t)$ by two cubic spline functions:

$$n(\xi,t) = c_0 + c_1\xi + c_2\xi^2 + c_3\xi^3, \quad (22)$$

where the coeficients c_i $(i = 0,\ldots,3)$ depend on the moments Q_0, Q_1, Q_2.

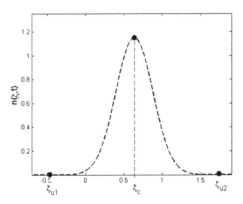

Fig. 4. Gaussian distribution of $n(\xi,t)$. Signed nodes ξ_c and ξ_u are used for determining an appropriate spline function

Each segment of the spline respects the maximum value (ξ_c) of the Gaussian distribution of $n(\xi,t)$ and the point (ξ_u) where $n(\xi,t)$ is close to zero. These values are displayed in fig. 4. The maximal value of $n(\xi,t)$ is attained for ($\xi_c = p$), where p is defined in (21). The coefficients c_i follow from conditions for values of function and their first derivatives in the end points of the interval (ξ_u, ξ_c):

$$\begin{bmatrix} 1 & \xi_c & \xi_c^2 & \xi_c^3 \\ 1 & \xi_u & \xi_u^2 & \xi_u^3 \\ 0 & 1 & 2\xi_c & 3\xi_c^2 \\ 0 & 1 & 2\xi_u & 3\xi_u^2 \end{bmatrix} \begin{bmatrix} c_0 \\ c_1 \\ c_2 \\ c_3 \end{bmatrix} = \begin{bmatrix} n(\xi_c) \\ 0 \\ 0 \\ 0 \end{bmatrix}. \quad (23)$$

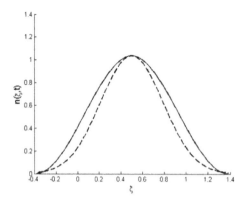

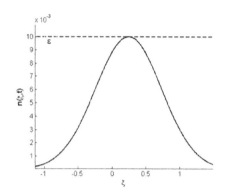

Fig. 5. Comparison of Gaussian approximation (dashed lines) to spline approximation (solid lines) of $n(\xi, t)$

Fig. 6. Gauss. distrib. of $n(\xi, t)$ for small Q_i (solid line) with the limit ε (dashed line)

Solving (23) gives:

$$c_0 = \frac{Q_0}{\sqrt{2\pi}} \frac{\xi_u^2(3\xi_c - \xi_u)}{(\xi_c - \xi_u)^3 \sqrt{Q_2/Q_0 - (Q_1/Q0)^2}}, \tag{24}$$

$$c_1 = -\frac{Q_0 3\sqrt{2}}{\sqrt{\pi}} \frac{\xi_c \xi_u}{(\xi_c - \xi_u)^3 \sqrt{Q_2/Q_0 - (Q_1/Q0)^2}}, \tag{25}$$

$$c_2 = \frac{Q_0 3}{\sqrt{2\pi}} \frac{\xi_c + \xi_u}{(\xi_c - \xi_u)^3 \sqrt{Q_2/Q_0 - (Q_1/Q0)^2}}, \tag{26}$$

$$c_3 = \frac{Q_0\sqrt{2}}{\sqrt{2\pi}} \frac{1}{(\xi_c - \xi_u)^3 \sqrt{Q_2/Q_0 - (Q_1/Q0)^2}}, \tag{27}$$

where the limit value ξ_u fulfills the assumption:

$$\xi_u : \qquad n(\xi_u, t) < \varepsilon \ll 1. \tag{28}$$

Solving (28) we obtain the expression for ξ_u which is a function of distribution moments:

$$\xi_{u1} > p - q\sqrt{2}\sqrt{-\ln\frac{\varepsilon}{a_0}}, \qquad \xi_{u2} < p + q\sqrt{2}\sqrt{-\ln\frac{\varepsilon}{a_0}}, \tag{29}$$

with $a_0 = \frac{Q_0}{q\sqrt{2\pi}}$ and $\xi_u = \xi_{u1}$ corresponding to the left spline branch and $\xi_u = \xi_{u2}$ to the right spline branch. Fig. 5 demonstrates the comparison of the Gaussian approximation and the spline approximation of $n(\xi, t)$. Moreover (21) and (28) give rise to four conditions of solvability:

$$\frac{Q_0}{\sqrt{2\pi}q} > 0, \qquad q > 0, \qquad Q_0 Q_2 \leq Q_1^2 \frac{Q_0}{\sqrt{2\pi}q} < \epsilon. \tag{30}$$

Note that the conditions (30) do not limit the possible solutions. They correspond to physical principles: fiber stiffness and generated force are greater than zero. The first three conditions hold also for the Gaussian approximation, which is fitted to the experiment, the last one arises from the spline approximation. If the last condition does not hold, it means that all Q_i are small and the situation displayed in fig. 6 occurs. In such a case it is possible to decrease ε, so that this condition does not resctrict the solution.

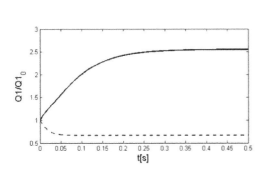
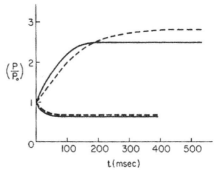

Fig. 7. Comparison of stretch and release response of the original Huxley model (solid lines) and the DM approximation (dashed lines) with the Gaussian approximation of $n(\xi,t)$ (right, taken from [19]) and the DM approximation with the spline approximation of $n(\xi,t)$ (left), constant velocity $w = \pm 10\,\mathrm{s}^{-1}$.

The behavior of the selected spline approximation was tested for correctness on a simple example and it was compared to a test published in [19]. The results are concluded by the graphs in fig. 7, with the first distribution moment displayed on the vertical axis normalized by its initial condition. For this comparison the initial conditions and parameters were set the same as in [19].

3.7. Equilibrium of active and passive elements

The passive properties of the muscle tissue, which are associated to the connective tissue components, are simulated by a viscoelastic component in series with the contractile element [17], recall fig. 3. The total stretch of the active fiber is the summation of its two components:

$$\lambda = \frac{l}{L} = c_{CE}\lambda_{CE} + c_{VE}\lambda_{VE} = c_{CE}\frac{l_{CE}}{L_{CE}} + c_{VE}\frac{l_{VE}}{L_{VE}}, \tag{31}$$

where l_{CE}, l_{VE} denote the instantaneous and L_{CE}, L_{VE} the rest lengths of particular components with the coefficients c_{CE}, c_{VE}. According to [17] it is assumed that w is proportional to λ_{CE}. It is necessary to note that w is defined as contraction while λ_{CE} represents elongation, therefore

$$w = -c_w \dot{\lambda}_{CE}, \tag{32}$$

with c_w a proportional coefficient.

The stress response of the VE obeys:

$$\tau(t, \lambda_{VE}) = (1-\gamma)\sigma(\lambda_{VE}(t)) + \gamma \int_0^t e^{-(t-\theta)/T} \dot{\sigma}(\lambda_{VE}(\theta))\,\mathrm{d}\theta, \tag{33}$$

$$\sigma(\lambda_{VE}) = E(\lambda_{VE} - 1) = E\left(\frac{\lambda - c_{CE}\lambda_{CE}}{c_{VE}} - 1\right), \tag{34}$$

where σ_{CE} is the linear elastic response with E the Young modulus, γ is the relaxation parameter and T the relaxation time.

As CE and VE are connected in series, they transmit the same tension and the whole element is in the equilibrium:

$$\tau(\lambda, \lambda_{CE}, t) = \sigma_{CE}(\lambda_{CE}, t). \tag{35}$$

If the total deformation λ is given then λ_{CE} can be computed from the equilibrium equation (35) and λ_{VE} satisfies (31).

4. Implementation of the complete model into MAFEST

The material model of the active skeletal muscle fiber described in the previous sections was implemented into MAFEST. This software is based on the combination of two programming languages: Matlab and C, cf. [2]. The program core containing material definitions and FE assembling is written in C. The user interface and the program logic are implemented in Matlab.

Recall that our material is a composite mixture (CM) consisting of three plys: a hyperelastic matrix, passive fibers and active fibers where, according to the theory of mixtures, the total stress of the material is the sum of stresses of particular plys weighted by their volume fractions as stated in (2). All plys share a common displacement field and no mutual movement is considered, see [2], where the constitutive relations for matrix and passive fibers are described in detail. The active fiber stress is defined by (16).

We assume our problem is discretized in space by finite elements (FE) [2]. In the TLF, all quantities and integrals are related to the initial configuration which is undeformed and the space derivatives are with respect to the material coordinates $\underline{X}$. The displacements $\underline{u}$ represent the unknown field, $\underline{x} = \underline{X} + \underline{u}$, $\underline{x} = \boldsymbol{\chi}^T \cdot \boldsymbol{x}$, $\underline{u} = \boldsymbol{\chi}^T \cdot \boldsymbol{u}$ and $\boldsymbol{\chi}$ are the FE base functions. In particular the tri-linear base functions are used in the numerical examples below with the hexahedral finite elements.

The integral in the weak form of the equilibrium equation (4) is evaluated numerically over each finite element using a numerical (Gauss) quadrature. Thus all the material parameters must be given in all the quadrature points in the whole domain. In MAFEST, any parameter, in particular the fibre directions, can be defined independently in each quadrature point, allowing for the inhomogeneity and anisotropy of the tissue. The internal equilibrium of the active fiber is solved in each quadrature point separately too.

The equilibrium equation of the CM is expressed below by the function $\boldsymbol{f}$ (discrete counterpart of (4)) and h corresponds to the internal equilibrium of the active fiber (discrete counterpart of (35)):

$$\boldsymbol{f}(\dot{\boldsymbol{u}}, \boldsymbol{u}, \boldsymbol{\lambda}_{CE}) = 0, \qquad \boldsymbol{u} = \{u_j\}_1^{n_u}, \tag{36}$$

$$h(\boldsymbol{u}, \boldsymbol{\lambda}_{CE}) = 0, \qquad \boldsymbol{\lambda}_{CE} = \{\lambda_{CEj}\}_1^{n_Q}, \tag{37}$$

where n_u is the number of displacement degrees of freedom per element, n_Q is the number of quadrature points per element and λ_{CE} is the internal variable in quadrature points.

The discretized second Piola-Kirchhoff stress tensor $\underline{S}$ in vector form is denoted by $\boldsymbol{s}$. Then in the FE discretization the term $\delta_u \underline{\underline{E}}(\underline{u}; \underline{v})$ turns into

$$\delta_u \underline{\underline{E}}(\underline{u}; \underline{v}) \approx \boldsymbol{B}(\boldsymbol{u}) \cdot \boldsymbol{v}, \tag{38}$$

and the equations (36), (37) can be written in the form

$$\boldsymbol{f}(\boldsymbol{u}, \boldsymbol{\lambda}_{CE}) := \sum_{q=1}^{n_Q} [\boldsymbol{B}(\boldsymbol{u})^T \cdot \boldsymbol{s}(\boldsymbol{u}, \boldsymbol{\lambda}_{CE}) J_0] = 0, \tag{39}$$

$$h_{|q} := \sigma_{CE}(\lambda_{CE}) - \tau(\lambda, \lambda_{CE}) = 0, \qquad q = 1 \ldots n_Q, \tag{40}$$

where the total stress $\boldsymbol{s}$ for $q = 1 \ldots n_Q$ (index omitted) is

$$\boldsymbol{s}(\boldsymbol{u}, \boldsymbol{\lambda}_{CE}) := \phi^m \boldsymbol{s}^m(\boldsymbol{u}) + \phi^a \tau^a(\boldsymbol{u}, \boldsymbol{\lambda}_{CE}) \boldsymbol{\omega} + \phi^p \tau^p(\boldsymbol{u}) \boldsymbol{\omega}. \tag{41}$$

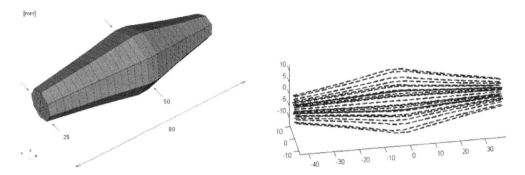

Fig. 8. Muscle bundle in the undeformed state (left), fiber orientation (right)

Assuming the time discretization, for the step $n + 1$ we obtain:

$$f(u^{(n+1)}, \lambda_{CE}^{(n+1)}) := \sum_{q=1}^{n_Q}[B(u^{(n+1)})^T \cdot s(u^{(n+1)}, \lambda_{CE}^{(n+1)})J_0]_{|\xi^q} = 0, \qquad (42)$$

$$h_{|q} := \sigma_{CE}(\lambda_{CE}^{(n+1)}) - \tau(\lambda^{(n+1)}, \lambda_{CE}^{(n+1)}) = 0, \qquad q = 1 \ldots n_Q, \qquad (43)$$

where the discretized equation for defining $h_{|q}$ is itemized in detail in [12]. The nonlinear problem (36)–(37) is solved by the Newton method. The above equations, related to a single finite element, are assembled for the whole discrete domain in the usual finite element sense.

5. Examples

The validation of the active fiber material model according to tests presented in literature can be found in [11] or [12]. In our case the whole implemented skeletal muscle model is tested on a simple muscle-like geometry. The shape reminding a muscle bundle is depicted undeformed in fig. 8, left. Both the active and passive fibers are defined in the longitudinal direction as is shown in fig. 8, right. The small model dimensions are chosen due to fact that all parameters found in literature [6, 7, 19] are measured on small specimens (frog or rats). The used parameters of active fibers are summarized in Tab. 1, where f_1, g_1, g_2 and g_3 belong to the attachment and detachment rates $f(\xi)$, $g(\xi)$.

Table 1. Active fiber parameters

Contractile element							Viscoelatic element		
f_1 [s^{-1}]	g_1 [s^{-1}]	g_2 [s^{-1}]	g_3 [s^{-1}]	θ [s^{-1}]	μ [–]	K_A [kPa]	γ [–]	E [kPa]	T [s^{-1}]
35	7	200	30	11.25	0.2	400	0.75	10^4	0.1

The isometric contraction of the muscle model lasting 0.5 s is displayed in fig. 9 (a). The reaction to the external loading which is lower (4 kPa) than the maximal force generated by the muscle, is denoted as the concentric contraction and can be seen in fig. 9 (b). While the excentric contraction occurs when the external loading is greater than the maximal force generated by the muscle (20 kPa) and is presented in fig. 9 (c).

In all previous simulations the muscle model swells in its middle part. It is caused by the uniform distribution of the active fiber parameter K_A. This parameter is related to the fiber

Fig. 9. Isometric (a), concentric(b) and excentric (c) contraction; constant K_A

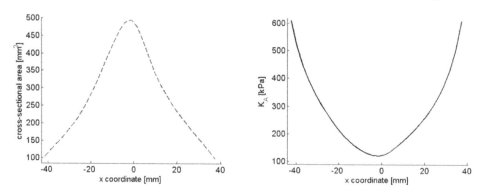

Fig. 10. Cross-section area along the bundle (left), distribution of K_A for const $= 60\,000$

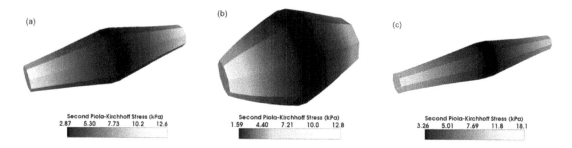

Fig. 11. Isometric (a), concentric and excentric (c) contraction; $K_A = K_A(x)$

cross-section, which causes that the greater cross-sectional area contains more active fibers. In consequence the active force generated in the middle of the bundle is greater due to this greater cross-sectional area. However according to anatomy sources a muscle bundle contains a constant number of fibers in an arbitrary cross-section, i.e. any muscle bundle generates a constant force through its cross sections. Thus we define a K_A as a function $K_A = K_A(x)$, such that the generated force remains constant and x changes along the muscle:

$$K_A(x) = \frac{\text{const}}{S(x)}, \tag{44}$$

where $S(x)$ is the cross-sectional area dependent on the longitudinal coordinate x, see fig. 10.

The previous examples are now recomputed with K_A dependent on the cross-sectional area. As depicted in fig. 11 (a), the isometric contraction is now without the "belly effect", the strains are very small. For the comparison see also fig. 11 (b) and (c) where the concentric and excentric contractions are displayed.

6. Conclusion

The paper deals with the 3D skeletal muscle modeling. The muscle tissue is formed by a matrix, passive fibers and active fibers. The implementation of the model uses the constitutive relations provided by MAFEST for the passive fiber and the matrix, and for the active fibers the new material model.

The active fiber behavior is based on the Huxley type model respecting the tissue microstructure and involving the calcium activation. The fiber microstructure is represented by a function describing the actual distribution of bonded cross-bridges. The new approximation of the cross-bridge distribution function leading to the easier computer implementation was proposed.

The simple muscle bundle-like geometry was created and the isometric concentric and excentric contractions of our 3D muscle model were shown. The presented model is able to simulate various situations during muscular contraction. It is also possible to apply the model to an arbitrary geometry and respect the fiber architecture. It is suitable for modeling of isolated muscles. However the implementation into a whole body model is not effective because of its high computational complexity.

Acknowledgements

The work has been supported by John H. and Anny Bowles Foundation.

References

[1] Bannister, L. H., Berry, M. M., Collins, P., Dussek, J. E., Dyson, M., Gray's Anatomy, Churchil Livingstone, Great Britain, 1995.

[2] Cimrman, R., Mathematical modeling of biological tissues, Ph.D. thesis, University of West Bohemia, Pilsen, 2002.

[3] Cimrman, R., Rohan, E., Modelling heart tissue using a composite muscle model with blood perfusion, Proceedings of Computational Fluid and Solids Mechanics 2003, Elsevier Science Ltd., 2003.

[4] Cimrman, R., Rohan, E., Nováček, V., Mechanical modelling and parameter identification of soft tissue: Kidney case study, Proceedings of Computational Mechanics 2005, Hrad Nečtiny, University of West Bohemia, Pilsen, 2005, pp. 111–118.

[5] Cimrman, R., Rohan, E., On identification of the arterial model parameters from experiments applicable "in vivo", Mathematics and Computers in Simulation (2009), in press.

[6] Gielen, A. W. J., A continuum approach to the mechanics of contracting skeletal muscle, Ph.D. thesis, Eindhoven University of Technology, 1998.

[7] Hatze, H., Myocybernetic control models of skeletal muscle: Characteristics and applications, University of South Africa, Muckleneuk, Pretoria 1981.

[8] Holtzapfel, G. A., Nonlinear Solid Mechanics, John Wiley & Sons Ltd., England, 2000.

[9] Huxley, A. F., The mechanism of muscular contraction, Science 164 (1969) 1 356–1 366.

[10] Keener, J., Sneyd, J., Mathematical Physiology, Springer-Verlag, New York, 1998.

[11] Kocková, H., Cimrman, R., On skeletal muscle model with new approximation of cross-bridge distribution, Proceedings of Computational mechanics, Hrad Nečtiny, University of West Bohemia, 2006, pp. 255–260.

[12] Kocková, H., Biomechanical models of living tissues and their industrial applications, Ph.D. thesis, University of West Bohemia, Pilsen, 2007.

[13] Kochová, P., Cimrman., R., Rohan, E., Parameter identification and mechanical modelling of smooth muscle and connective tissue, Proceedings of Applied mechanics 2007, Technical University of Ostrava, Ostrava, 2007, pp. 133–135.

[14] Kochová, P., Cimrman, R., Rohan, E., Orientation of smooth muscle cells with application in mechanical model of gastropod tissue, Applied and Computational Mechanics 3, (2009), in press.

[15] Oomens, C. W. J., Maenhout, M., van Oijen, C. H., Drost, M. R., Baaijens, F. P., Finite element modelling of contracting skeletal muscle, The Royal Society 358 (2003) 1 453–1 460.

[16] Rohan, E., Cimrman, R., Numerical simulation of activated smooth muscle behaviour using finite elements, Proceedings of UWB, University of West Bohemia, Pilsen, 1999, pp. 143–155.

[17] Rohan, E., On coupling the sliding cross-bridge model of muscle with series viscoelastic element, Proceedings of Computational Mechanics, Hrad Nečtiny, University of West Bohemia, Pilsen, 2002, pp. 395–402.

[18] Sainte-Marie, J., Chapelle, D., Cimrman, R., Sorine, M., Modeling and estimation of the cardiac electromechanical activity, Computers and Structures 84 (2006) 1 743–1 759.

[19] Zahalak, G. I., A distributed moment approximation for kinetic theories of muscular contraction, Mathematical Biosciences 55 (1981) 89–114.

[20] Zahalak, G. I., Ma, S. P., Muscle activation contraction: constitutive relations based directly on cross-bridge kinetics, Journal of Biomechanical Engineering 112 (1990) 52–62.

[21] World of the Body: skeletal muscle, http://www.answers.com/topic/skeletal-muscle.

16

Numerical approximation of flow in a symmetric channel with vibrating walls

P. Sváček[a,b,*], J. Horáček[a]

[a]Institute of Thermomechanics, Czech Academy of Sciences, Dolejškova 5, 182 00 Praha 8, Czech Republic

[b]Dep. of Technical Mathematics, Faculty of Mechanical Engineering, Czech Technical University in Prague, Karlovo nam. 13, 121 35 Praha 2, Czech Republic

Abstract

In this paper the numerical solution of two dimensional fluid-structure interaction problem is addressed. The fluid motion is modelled by the incompressible unsteady Navier-Stokes equations. The spatial discretization by stabilized finite element method is used. The motion of the computational domain is treated with the aid of Arbitrary Lagrangian Eulerian (ALE) method. The time-space problem is solved with the aid of multigrid method.

The method is applied onto a problem of interaction of channel flow with moving walls, which models the air flow in the glottal region of the human vocal tract. The pressure boundary conditions and the effects of the isotropic and anisotropic mesh refinement are discussed. The numerical results are presented.

Keywords: biomechanics of voice production, incompressible Navier-Stokes equations, finite element method, Arbitrary Lagrangian-Eulerian method

1. Introduction

This paper is concerned with numerical simulation of unsteady viscous incompressible flow in a simplified model of the glottal region of the human vocal tract with the aid of the finite element method (FEM). The main attention is paid to the efficient computation of the flow field. For the robust and efficient solver both the advanced stabilization (as streamline upwind/Petrov Galerkin stabilizations, cf. [6, 7]) and solution methods (as multigrid and/or domain decomposition, cf. [19, 9, 10, 13]) have to be employed.

FEM is well known as a general discretization method for partial differential equations. It can handle easily complex geometries and also boundary conditions employing derivatives. However, straightforward application of FEM procedures often fails in the case of incompressible Navier-Stokes equations. The reason is that momentum equations are of advection-diffusion type with dominating advection. The Galerkin FEM leads to unphysical solutions if the grid is not fine enough in regions of strong gradients (e.g. boundary layer). In order to obtain physically admissible correct solutions it is necessary to apply suitable mesh refinement (e.g. anisotropically refine mesh, cf. [5]) combined with a stabilization technique, cf. [7, 3, 18, 16].

Furthermore, the time and space discretized linearized problem of the arising large system of linear equations needs to be solved in fast and efficient manner. The application of direct solvers as UMFPACK (cf. [4]) leads to robust method, where different stabilizations procedures can be easily applied even on anisotropically refined grids. However, the application of direct solver

---

*Corresponding author. e-mail: Petr.Svacek@fs.cvut.cz.

for system of equations with more than approximately 10^5 unknowns becomes unfeasible in many cases (depending on computer CPU and memory).

In that case the application of multigrid (cf. [19]) or domain decomposition methods is an option, cf. [13]. In this paper a simplified version of multigrid method is shortly described together with a choice of finite elements and stabilization procedures. Even when the method is simplified, it was found to be efficient and robust enough.

The developed method is applied to the numerical solution of a channel flow modelling the glottal region of the human vocal tract including the vibrating vocal fold. The vibrations of the channel wall are prescribed, see [14]. Further, in order to obtain physically relevant results the pressure drop boundary conditions are employed, cf. [8].

First, the mathematical model consisting of time dependent computational domain and incompressible flow model is introduced. Further, in Section 3 the time and space discretization is described and Section 4 describes the application of a simple multigrid version. Section 5 shows the numerical results.

2. Mathematical model

The model problem consists of flow model, which describes the fluid motion in the time-dependent computational domain Ω_t, i.e. in a channel with moving walls, see Fig. 1. For the description and the approximation on moving meshes the Arbitrary Lagrangian-Eulerian (ALE) method is employed, cf. [12]. The geometry of the channel is chosen according [14], where a different distance between the moving walls, i.e. the gap $g(t)$, was considered. Further, on the outlet part of the channel a modification of do-nothing boundary condition was applied in order to allow the vortices flow smoothly out of the computational domain. On the inlet either the Dirichlet boundary condition for velocity is prescribed or preferably we use the pressure drop formulation, similarly as in cf. [8]. The presented mathematical model (and also its numerical approximation) is a slight modification of the mathematical model applied to the numerical simulation of flow induced airfoil vibrations in our previous works, cf. [18].

2.1. Arbitrary Lagrangian Eulerian method

In order to treat the fluid flow on moving domains, the so-called Arbitrary Lagrangian Eulerian method is used. We assume that $\mathcal{A} = \mathcal{A}(\xi, t) = \mathcal{A}_t(\xi)$ is an ALE mapping defined for all

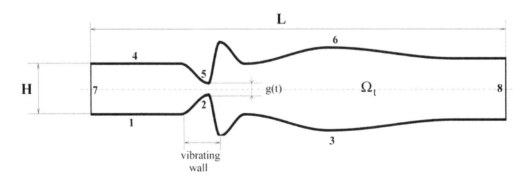

Fig. 1. Computational domain and boundary parts: The inlet part of the boundary Γ_I (number 7), the outlet part of the boundary Γ_O (number 8), the fixed walls Γ_D (numbers 1,4,3,6) and vibrating walls Γ_{Wt} (numbers 2, 5)

$t \in (0,T)$ and $\xi \in \Omega_0$, which is smooth enough and continuously differentiable mapping of $\xi \in \Omega_0$ onto $x \in \Omega_t$, $x = \mathcal{A}_t(\xi)$. We define the *domain velocity* $\mathbf{w}_D : \mathcal{M} \to \mathbb{R}$ satisfies

$$\mathbf{w}_D(x,t) = \frac{\partial \mathcal{A}}{\partial t}(\xi,t) \qquad \text{for all } \xi \in \Omega_0 \text{ and } t \in (0,T), \quad \text{where } x = \mathcal{A}(\xi,t). \tag{1}$$

Furthermore the symbol $D^{\mathcal{A}}/Dt$ denotes the ALE derivative, i.e. the time derivative with respect to the reference configuration. The ALE derivative satisfies (cf. [18, 11])

$$\frac{D^{\mathcal{A}} f}{Dt}(x,t) = \frac{\partial f}{\partial t}(x,t) + \mathbf{w}_D(x,t) \cdot \nabla f(x,t). \tag{2}$$

In the present paper the ALE mapping is analytically prescribed, but in the future this mapping will be a part of the solution similarly as in cf. [18].

2.2. Flow model

Let us consider the following system of the incompressible Navier-Stokes equations in a bounded time-dependent domain $\Omega_t \subset R^2$ written in ALE form

$$\frac{D^{\mathcal{A}} \mathbf{v}}{Dt} - \nu \triangle \mathbf{v} + ((\mathbf{v} - \mathbf{w}_D) \cdot \nabla)\mathbf{v} + \nabla p = 0, \qquad \nabla \cdot \mathbf{v} = 0, \qquad \text{in } \Omega_t, \tag{3}$$

where $\mathbf{v} = \mathbf{v}(x,t)$ is the flow velocity, $p = p(x,t)$ is the kinematic pressure (i.e. pressure divided by the constant fluid density ρ_∞) and ν is the kinematic viscosity.

The boundary of the computational domain $\partial \Omega_t$ consists of mutually disjoint parts Γ_D (wall), Γ_I (inlet), Γ_O (outlet) and the moving part Γ_{Wt} (oscillating wall). The following boundary conditions are prescribed

$$
\begin{aligned}
&\text{a)} \quad \mathbf{v}(x,t) = \mathbf{0} && \text{for } x \in \Gamma_D, \\
&\text{b)} \quad \mathbf{v}(x,t) = \mathbf{w}_D(x,t) && \text{for } x \in \Gamma_{Wt}, \\
&\text{c)} \quad -(p - p_{ref}^o)\mathbf{n} + \tfrac{1}{2}(\mathbf{v} \cdot \mathbf{n})^- \mathbf{v} + \nu \frac{\partial \mathbf{v}}{\partial \mathbf{n}} = 0, && \text{on } \Gamma_O, \\
&\text{d)} \quad -(p - p_{ref}^i)\mathbf{n} + \tfrac{1}{2}(\mathbf{v} \cdot \mathbf{n})^- \mathbf{v} + \nu \frac{\partial \mathbf{v}}{\partial \mathbf{n}} = 0, && \text{on } \Gamma_I,
\end{aligned}
\tag{4}
$$

where $\mathbf{n}$ denotes the unit outward normal vector, the constants p_{ref}^i, p_{ref}^o denotes the reference pressure values, and α^- denotes the negative part of a real number α. In computations the condition (4d) can be replaced by the condition

$$\text{e)} \quad \mathbf{v}(x,t) = \mathbf{v}_D \qquad \text{for } x \in \Gamma_I. \tag{5}$$

Finally, we prescribe the initial condition

$$\mathbf{v}(x,0) = \mathbf{v}^0(x) \qquad \text{for } x \in \Omega_0.$$

3. Numerical approximation

In this section the numerical approximation of the mathematical model given in Section 2 is shown. As already mentioned the presented numerical approximation is a slight modification of our previous works, cf. [18, 17]. Nevertheless there are several significant differences, which were found to be important for the numerical approximation: boundary conditions used on the inlet/outlet part of the computational domain and their weak formulation, a modified Galerkin/Least-Squares (GLS) scheme employed for stable pair of finite elements, and the choice of stabilizing parameters. The space discretization and its stabilization is briefly desribed for the sake of clarity and completeness.

3.1. Time discretization

We consider a partition $0 = t_0 < t_1 < \ldots < T, t_k = k\Delta t$, with a time step $\Delta t > 0$, of the time interval $(0, T)$ and approximate the solution $\mathbf{v}(\cdot, t_n)$ and $p(\cdot, t_n)$ (defined in Ω_{t_n}) at time t_n by $\mathbf{v}^n$ and p^n, respectively. For the time discretization we employ a second-order two-step scheme using the computed approximate solution $\mathbf{v}^{n-1}$ in $\Omega_{t_{n-1}}$ and $\mathbf{v}^n$ in Ω_{t_n} for the calculation of $\mathbf{v}^{n+1}$ in the domain $\Omega_{t_{n+1}} = \Omega_{n+1}$. We write

$$\frac{\partial \mathbf{v}}{\partial t}(x, t^{n+1}) \approx \frac{3\mathbf{v}^{n+1} - 4\hat{\mathbf{v}}^n + \hat{\mathbf{v}}^{n-1}}{2\Delta t} \qquad \text{where } x \in \Omega_{n+1}, \tag{6}$$

where $\hat{\mathbf{v}}^n$ and $\hat{\mathbf{v}}^{n-1}$ are the approximate solutions $\mathbf{v}^n$ and $\mathbf{v}^{n-1}$ defined on Ω_n and Ω_{n-1}, respectively, and transformed onto Ω_{n+1} with the aid of ALE mapping, i.e. $\hat{v}^i(x) = \mathbf{v}^i(\mathcal{A}_{t_i}(\xi))$ where $x = \mathcal{A}_{t_{n+1}}(\xi) \in \Omega_{n+1}$. Further, we approximate the domain velocity $\mathbf{w}_D(x, t_{n+1})$ by $\mathbf{w}_D{}^{n+1}$, where

$$\mathbf{w}_D{}^{n+1}(x) = \frac{3\mathcal{A}_{t_{n+1}}(\xi) - 4\mathcal{A}_{t_n}(\xi) + \mathcal{A}_{t_{n-1}}(\xi)}{2\Delta t}, \qquad x = \mathcal{A}_{t_{n+1}}(\xi), \ x \in \Omega_{n+1}.$$

Then the time discretization leads to the following problem in domain Ω_{n+1}

$$\frac{3\mathbf{v}^{n+1} - 4\hat{\mathbf{v}}^n + \hat{\mathbf{v}}^{n-1}}{2\Delta t} - \nu \triangle \mathbf{v}^{n+1} + \left((\mathbf{v}^{n+1} - \mathbf{w}_D{}^{n+1}) \cdot \nabla\right) \mathbf{v}^{n+1} + \nabla p^{n+1} = 0, \tag{7}$$

$$\nabla \cdot \mathbf{v}^{n+1} = 0,$$

equipped with boundary conditions (4a–d) and the initial condition.

3.2. Weak formulation

For solution of the problem by finite element method, the time-discretized problem (7) is reformulated in a weak sense. The following notation is used: By $W = \mathbf{H}^1(\Omega_{n+1})$ the velocity space is defined, by X the space of test functions is denoted

$$X = \{\mathbf{z} \in W : \mathbf{z} = 0 \text{ on } \Gamma_{Wt_{n+1}} \cup \Gamma_D\},$$

and by $Q = L^2(\Omega_{n+1})$ the pressure space is denoted. Using the standard approach, cf. [18], the solution $\mathbf{v} = \mathbf{v}^{n+1}$ and $p = p^{n+1}$ of problem (7) satisfies

$$a(U, V) = f(V), \qquad U = (\mathbf{v}, p) \tag{8}$$

for any $V = (\mathbf{z}, q) \in X \times Q$. Here, the forms $a(\cdot, \cdot)$ and $f(\cdot)$ are defined for any U, V by

$$a(U, V) = \left(\frac{3}{2\Delta t}\mathbf{v}, \mathbf{z}\right) + \nu\left(\nabla\mathbf{v}, \nabla\mathbf{z}\right) + \mathcal{B}(\mathbf{v}, \mathbf{z}) + c_n(\mathbf{v}; \mathbf{v}, \mathbf{z}) - (p, \nabla \cdot \mathbf{z}) + (\nabla \cdot \mathbf{v}, q),$$

$$f(V) = \frac{1}{2\Delta t}\left(4\hat{\mathbf{v}}^n - \hat{\mathbf{v}}^{n-1}, \mathbf{z}\right) - \int_{\Gamma_I} p_{ref}^i \mathbf{v} \cdot \mathbf{n}\, dS - \int_{\Gamma_O} p_{ref}^o \mathbf{v} \cdot \mathbf{n}\, dS,$$

and for any $\mathbf{w}, \mathbf{v}, \mathbf{z} \in W$

$$c_n(\mathbf{w}, \mathbf{v}, \mathbf{z}) = \int_{\Omega_{n+1}} \left(\frac{1}{2}(\mathbf{w} \cdot \nabla\mathbf{v}) \cdot \mathbf{z} - \frac{1}{2}(\mathbf{w} \cdot \nabla\mathbf{z}) \cdot \mathbf{v}\right)\, dx - \left((\mathbf{w}_D{}^{n+1} \cdot \nabla)\mathbf{v}, \mathbf{z}\right),$$

$$\mathcal{B}(\mathbf{v}, \mathbf{z}) = \int_{\Gamma_I \cup \Gamma_O} \frac{1}{2}(\mathbf{v} \cdot \mathbf{n})^+ \mathbf{v} \cdot \mathbf{z}\, dS,$$

where by $(\cdot, \cdot)$ we denote the scalar product in the space $L^2(\Omega_{n+1})$.

3.3. Spatial discretization

Further, the weak formulation (8) is approximated by the use of FEM: we restrict the couple of spaces (X, M) to finite element spaces (X_h, M_h). First, the computational domain Ω_t is assumed to be polygonal and approximated by an admissible triangulation $\mathcal{T}_h$, cf. [2]. Based on the triangulation $\mathcal{T}_h$ the Taylor-Hood finite elements are used, i.e.

$$\mathcal{H}_h = \{v \in C(\overline{\Omega_{n+1}}); v|_K \in P_2(K) \text{ for each } K \in \mathcal{T}_h\},$$
$$\mathcal{W}_h = [\mathcal{H}_h]^2, \qquad X_h = \mathcal{W}_h \cap \mathcal{X}, \qquad (9)$$
$$\mathcal{M}_h = \{v \in C(\overline{\Omega_{n+1}}); v|_K \in P_1(K) \text{ for each } K \in \mathcal{T}_h\}.$$

The couple (X_h, M_h) satisfy the Babuška-Brezzi inf-sup condition, which guarantees the stability of a scheme, cf. [20].

Problem 1 (Galerkin approximations). *Find* $U_h = (\mathbf{v}_h, p_h) \in (X_h, M_h)$ *such that* $\mathbf{v}_h$ *satisfy boundary conditions (4a,b) and*

$$a(U_h, V_h) = f(V_h), \qquad (10)$$

for all $\mathbf{z}_h \in X_h$ *and* $q_h \in M_h$.

The Galerkin approximations are unstable in the case of high Reynolds numbers, when the convection dominates. In that case a stabilized method needs to be applied.

3.4. Stabilization

In order to overcome the above mentioned instability of the scheme, modified Galerkin Least Squares method is applied, cf. [7]. We start with the definition of the local element rezidual terms $\mathcal{R}_K^a$ and $\mathcal{R}_K^f$ defined on the element $K \in \mathcal{T}_h$ by

$$\mathcal{R}_K^a(\tilde{\mathbf{w}}; \mathbf{v}, p) = \frac{3\mathbf{v}}{2\Delta t} - \nu\triangle\mathbf{v} + (\tilde{\mathbf{w}} \cdot \nabla)\mathbf{v} + \nabla p, \qquad \mathcal{R}_K^f(\hat{\mathbf{v}}^n, \hat{\mathbf{v}}^{n-1}) = \frac{4\hat{\mathbf{v}}^n - \hat{\mathbf{v}}^{n-1}}{2\Delta t}. \qquad (11)$$

Further, the stabilizing terms are defined for $U^* = (\mathbf{v}^*, p^*)$, $U = (\mathbf{v}, p)$, $V = (\mathbf{z}, q)$ by

$$\mathcal{L}_{GLS}(U^*; U, V) = \sum_{K \in \mathcal{T}_h} \delta_K \left(\mathcal{R}_K^a(\tilde{\mathbf{w}}; \mathbf{v}, p), (\tilde{\mathbf{w}} \cdot \nabla)\mathbf{z} + \nabla q \right)_K,$$
$$\mathcal{F}_{GLS}(V) = \sum_{K \in \mathcal{T}_h} \delta_K \left(\mathcal{R}_K^f(\hat{\mathbf{v}}_n, \hat{\mathbf{v}}_{n-1}), (\tilde{\mathbf{w}} \cdot \nabla)\mathbf{z} + \nabla q \right)_K, \qquad (12)$$

where the function $\tilde{\mathbf{w}}$ stands for the transport velocity, i.e. $\tilde{\mathbf{w}} = \mathbf{v}^* - \mathbf{w}_D^{n+1}$. The additional grad-div stabilization terms read

$$\mathcal{P}_h(U, V) = \sum_{K \in \mathcal{T}_h} \tau_K (\nabla \cdot \mathbf{v}, \nabla \cdot \mathbf{z})_K.$$

In the case of bounded convection velocity the choice of parameters according [7] for BB stable pair of FE (reduced scheme) would be possible. However, in order to obtain a fast and efficient multigrid method, the following choice of the parameters δ_K and τ_K is used

$$\tau_K = \nu \left(1 + Re^{loc} + \frac{h_K^2}{\nu \Delta t} \right), \qquad \delta_K = \frac{h_K^2}{\tau_K},$$

where the local Reynolds number Re^{loc} is defined as $Re^{loc} = \frac{h\|\mathbf{v}\|_K}{2\nu}$.

Problem 2 (Galerkin Least Squares stabilized approximations). *We define the discrete problem to find an approximate solution $U_h = (\mathbf{v}_h, p_h) \in \mathcal{W}_h \times \mathcal{Q}_h$ such that $\mathbf{v}_h$ satisfies approximately conditions (4a,b) and the identity*

$$a(U_h, V_h) + \mathcal{L}_{GLS}(U_h; U_h, V_h) + \mathcal{P}_h(U_h, V_h) = f(V_h) + \mathcal{F}_{GLS}(V_h), \tag{13}$$

for all $V_h = (\mathbf{z}_h, q_h) \in \mathcal{X}_h \times \mathcal{Q}_h$.

4. Multigrid solution of the linear system

The space-time discretized system (13) needs to be solved by some linearization scheme, e.g. by Oseen linearization procedure described e.g. in [18] or [19]. The solution of the linearized system (13) leads to the solution of a modified saddle point system

$$S\underline{\mathbf{v}} + B\underline{p} = \underline{f}, \qquad \tilde{B}^T \underline{\mathbf{v}} + \tilde{A}\underline{p} = 0, \tag{14}$$

where $\underline{\mathbf{v}}$ and $\underline{p}$ is the finite-dimensional representation of the finite element approximations of velocity and pressure, respectively. Let us mention that for the non-stabilized system (i.e. in the case of $\delta_K \equiv \tau_K \equiv 0$) we have $\tilde{A} = 0$ and $\tilde{B} = B$.

From the system of equations (14) the pressure degrees of freedoms can be formally eliminated by formally multiplying the first equation of (14) by $\tilde{B}^T S^{-1}$ from the left, i.e. we get the system of equations

$$\left(\tilde{B}^T S^{-1} B - \tilde{A}\right) \underline{p} = \tilde{B}^T S^{-1} \underline{f}, \tag{15}$$

or with notation $A_p = \tilde{B}^T S^{-1} B - \tilde{A}$ and $g = \tilde{B}^T S^{-1} \underline{f}$ we have

$$A_p \underline{p} = g,$$

which can be solved by the Richardson iterative method

$$\underline{p}^{(l+1)} = \underline{p}^{(l)} + C^{-1}(g - A_p \underline{p}^{(l)}), \tag{16}$$

where C is a suitable preconditioner, see e.g. [19]. Nevertheless the choice of the preconditioner C is complicated in the case of convection dominated flows and the convergence of the scheme (16) is in this case slow. Moreover the stabilizing terms also badly influences the convergence rates.

In many cases and for small number of unknowns, the system can be solved with the aid of a direct solver, which yields fast, efficient and robust scheme. We refer to direct solver UMFPACK, cf. [4], which in the cases studied by the authors up to now [18] was efficient for number of unknowns less then approximately 10^5. However, with further increase of the number of unknowns the memory and CPU requirements grows too fast, so that the fast and efficient solution becomes impossible. One possibility is to use the parallel implementation of multi-frontal method, cf. [1].

Here, the solution of the system (14) is carried out by a simplified version of multi-grid method. Only single mesh and two levels of solution (coarse and fine grid levels) are used. The fine grid is represented by the used higher order finite elements (here Taylor-Hood finite elements, i.e. P2/P1 approximations for velocity/pressure). The coarse grid is considered as lower order finite elements (i.e. equal order P1/P1 approximations for velocity/pressure). The solution on the coarse grid can be obtained with the aid of direct solver UMFPACK, which

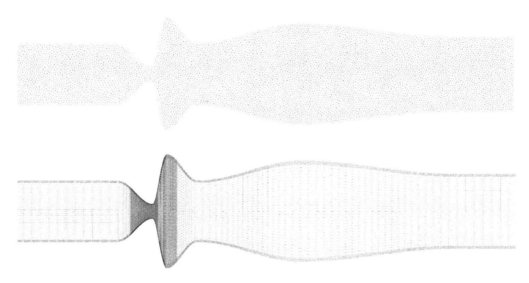

Fig. 2. The employed grids: the isotropic non-symmetric mesh (upper part) with $12\,219$ vertices and $23\,709$ elements and approximately 8×10^4 unknowns for flow problem, and the anisotropic axisymmetric mesh (lower part) with $8\,241$ vertices and $16\,000$ elements (resulting in 6×10^4 unknowns)

was found to be fast enough in the studied cases. On the fine grid the multiplicative Vanka-type smoother is used, cf. [9, 10]. This approach (i.e. the direct solver on coarse grid and Vanka-type smoother on fine grid) resulted in an efficient and fast method, which can be easily implemented. The performance of the multigrid method was found to be excellent for the isotropic grids. In the case of anisotropic mesh refinement, the convergence rates nevertheless become worse. The proper solution in this case is subject of a further study.

5. Numerical results

In this section the numerical results for air flow in a symmetric two-dimensional channel are presented. The channel geometry described in [14] is employed here, see also Fig. 1.

5.1. Stationary solution

First, we consider the non-moving computational domain Ω, where the influence of isotropically and anisotropically refined meshes is studied, see Fig. 2.

The following constants were used in the computations: fluid density $\rho_\infty = 1.225$ kg m^{-3} and kinematic viscosity $\nu = 1.5 \times 10^{-5}$ m^2/s, the width of the inlet part of the channel is $H = 0.0176$ m, the total length of the channel $L = 0.16$ m, and the constant gap width $g \equiv 4.4$ mm.

The boundary condition (4d) in the presented computations is replaced by the condition (5), where the constant flow velocity is presribed $\mathbf{v}_D(x,t) = (U_\infty, 0)^T$ at the inlet part of boundary Γ_I, and U_∞ was chosen in the range 0.01–0.05 m s^{-1}. The numerical results for stationary solution and different Reynolds numbers ($Re = \frac{1}{8}LU_\infty/\nu$) are presented in Figs. 3–4, where the isolines of the magnitude of velocity are shown. The results computed on both meshes for same Reynolds numbers show that even for low Reynolds numbers several stationary symmetric and nonsymmetric solutions exist. Fig. 3 (left) shows the symmetric solution obtained on both meshes for $Re = 20$. For $Re = 40$ and $Re = 50$ in Figs. 3–4 on isotropic mesh

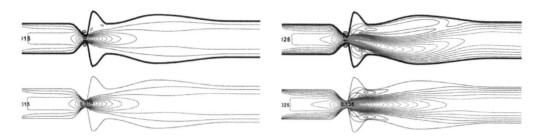

Fig. 3. The isolines of flow velocity magnitudes for Reynolds number 20 (left) and 40 (right) on isotropic mesh (upper part) and anisotropic mesh (lower part)

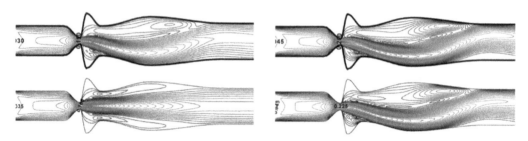

Fig. 4. The isolines of flow velocity magnitudes for Reynolds number 50 (left) and 70 (right) on isotropic mesh(upper part) and anisotropic mesh (lower part)

the non-symmetric solution was obtained, whereas on the anisotropical symmetric mesh the solution remains symmetric. For higher Reynolds number $Re > 50$ both solutions become non-symmetric.

5.2. Flow in channel with vibrating vocal folds

The numerical results for flow in vibrating channel are presented for physically relevant pressure drop, inlet flow velocity, frequency of vibrations and width of the channel, which leads to the Reynolds numbers in the range $Re = 1\,000 - 3\,000$.

The computations were carried out for the pressure drop of 400 Pa, i.e $p^i_{ref} = 400$ Pa and $p^o_{ref} = 0$ Pa. The initial condition was chosen as $\mathbf{v}^0 \equiv 0$ and the isotropically refined mesh was used, cf. Fig. 5. The gap oscillates harmonically around the mean gap value $\bar{g} = 4.4$ mm in the interval $g(t) \in [3.2 \text{ mm}, 5.6 \text{ mm}]$ with frequency $f = 100$ Hz .

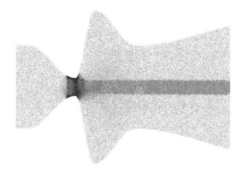

Fig. 5. The isotropic mesh used for multigrid solution of oscillating wall with 42 576 vertices and 84078 elements yielding approximately 4×10^5 unknowns for the flow problem

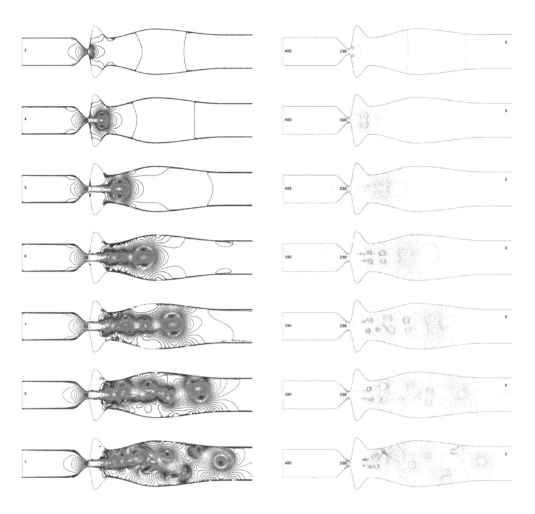

Fig. 6. The isolines of velocity magnitude (left) and pressure (right) in a sequence of time instance (from top to bottom, Part 1)

The results are shown in Figs. 6–7 for the time instants marked in Fig. 8. The sudden expansion in the modelled glottal region leads to the faster flow in the vibrating narrowest part of the computational domain and to complicated flow structures in the outlet part of the channel. Similar effects were observed experimentally in [15]. The inlet flow velocity and the flow velocity on the axis of symmetry at the narrowest part of the channel are shown in Fig. 8. The both values oscillates with a similar frequency as the prescribed motion of the wall. However, the graphs are noisy partially due to the complicated flow structures downstream.

6. Conclusion

The paper presents the developed mathematical method and applied numerical technique for solution of fluid-structure problems encountered in biomechanics of voice production. The method consists of the advanced stabilization of the finite element method applied considering the moving domain. In order to obtain fast solution of the discretized problem a simplified multigrid method was applied, which allowed solution of significantly larger system of equations compared to the previously used approach, see e.g. [18].

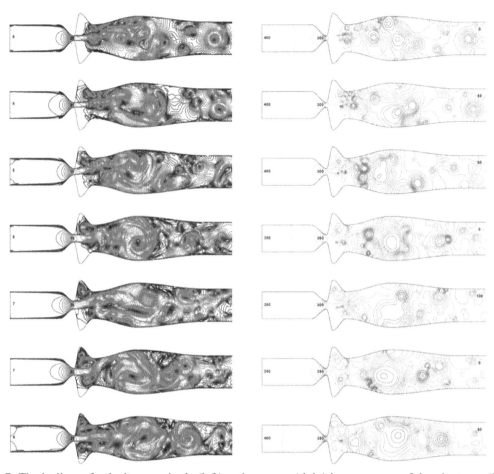

Fig. 7. The isolines of velocity magnitude (left) and pressure (right) in a sequence of time instance (from top to bottom, Part 2)

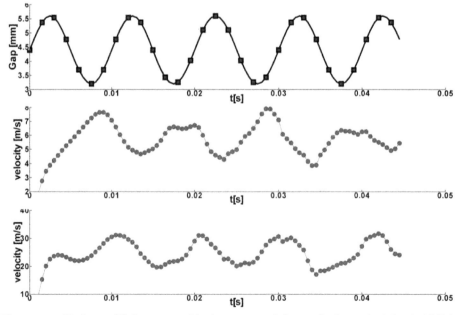

Fig. 8. The gap oscillations $g(t)$ (upper graph), the computed flow velocity at the inlet (middle), and the computed flow velocity in the glottal orifice (lower graph)

The influence of the isotropic and anisotropic meshes was studied and the multigrid technique was applied on a challenging problem of flow in symmetric channel with vibrating walls. The numerical results were presented showing the Coanda effect and complicated structure of small vortices and large size eddies generated at the glottal region by vibrating vocal fold. Similar vortex flow structures and Coanda effects were identified experimentally in [15].

Acknowledgements

This research was supported under the Project OC 09019 "Modelling of voice production based on biomechanics" within the program COST of the Ministry of Education of the Czech Republic and COST Action 1203 "Advanced Voice Function Assessment".

References

[1] Amestoy, P. R., Guermouche, A., L'Excellent, J.-Y., Pralet, S., Hybrid scheduling for the parallel solution of linear systems. Parallel Computing, 32:136–156, 2006.

[2] Ciarlet, P. G., The Finite Element Methods for Elliptic Problems. North-Holland Publishing, 1979.

[3] Codina, R., Stabilization of incompressibility and convection through orthogonal sub-scales in finite element methods. Computational Method in Applied Mechanical Engineering, 190:1 579–1 599, 2000.

[4] Davis, T. A., Duff, I. S., A combined unifrontal/multifrontal method for unsymmetric sparse matrices. ACM Transactions on Mathematical Software, 25:1–19, 1999.

[5] Dolejší, V., Anisotropic mesh adaptation technique for viscous flow simulation. East-West Journal of Numerical Mathematics, 9:1–24, 2001.

[6] Feistauer, M., Mathematical Methods in Fluid Dynamics. Longman Scientific & Technical, Harlow, 1993.

[7] Gelhard, T., Lube, G., Olshanskii, M. A., Starcke, J.-H., Stabilized finite element schemes with LBB-stable elements for incompressible flows. Journal of Computational and Applied Mathematics, 177:243–267, 2005.

[8] Heywood, J. G., Rannacher, R., Turek, S., Artificial boundaries and flux and pressure conditions for the incompressible Navier-Stokes equations. Int. J. Numer. Math. Fluids, 22:325–352, 1992.

[9] John, V., Higher order finite element methods and multigrid solvers in a benchmark problem for the 3D Navier-Stokes equations. Int. J. Num. Meth. Fluids, 40:775–798, 2002.

[10] John, V., Tobiska, L., Numerical performance of smoothers in coupled multigrid methods for the parallel solution of the incompressible Navier-Stokes equations. Int. J. Num. Meth. Fluids 33, 33:453–473, 2000.

[11] Nobile, F., Numerical approximation of fluid-structure interaction problems with application to haemodynamics. PhD thesis, Ecole Polytechnique Federale de Lausanne, 2001.

[12] Nomura, T., Hughes, T. J. R., An arbitrary Lagrangian-Eulerian finite element method for interaction of fluid and a rigid body. Computer Methods in Applied Mechanics and Engineering, 95:115–138, 1992.

[13] Otto, F. C., Lube, G., Non-overlapping domain decomposition applied to incompressible flow problems. Contemporary Mathematics, 218:507–514, 1998.

[14] Punčochářová, P., Fürst, J., Kozel, K., Horáček, J., Numerical simulation of compressible flow with low mach number through oscillating glottis. In J. Horáček, I. Zolotarev, editor, Flow Induced Vibrations, Prague, 2008. Institute of Thermomechanics, CAS.

[15] Šidlof, P., Fluid-structure interaction in human vocal folds. PhD thesis, Charles University, Faculty of Mathematics and Physics, 2007.

[16] Sváček, P., Feistauer, M., Application of a Stabilized FEM to Problems of Aeroelasticity. In Numerical Mathematics and Advanced Application, pp. 796–805, Berlin, 2004. Springer.

[17] Sváček, P., Feistauer, M., Horáček, J., Numerical simulation of a flow induced airfoil vibrations. In Proceedings Flow Induced Vibrations, volume 2, pp. 57–62, Paris, 2004. Ecole Polytechnique.

[18] Sváček, P., Feistauer, M., Horáček, J., Numerical simulation of flow induced airfoil vibrations with large amplitudes. Journal of Fluids and Structure, 23(3):391–411, 2007.

[19] Turek, S., Efficient Solvers for Incompressible Flow Problems: An Algorithmic and Computational Approach. Springer, Berlin, 1999.

[20] Verfürth, R., Error estimates for mixed finite element approximation of the Stokes equations. R.A.I.R.O. Analyse numérique/Numerical analysis, 18:175–182, 1984.

17

Validation of the adjusted strength criterion LaRC04 for uni-directional composite under combination of tension and pressure

J. Krystek[a,*], R. Kottner[a], L. Bek[a], V. Laš[a]

[a]*Faculty of Applied Sciences, University of West Bohemia, Univerzitní 22, 306 14 Plzeň, Czech Republic*

Abstract

Strength of unidirectional composite materials for some combinations of state of stress cannot be successsfully predicted even with modern failure criteria. In case of the combination of compression in the transverse direction and tension in the fiber direction, the criterion LaRC04 was adjusted in previous work. The predicted strengths in this case reach significantly larger values compared to the ultimate strengths of the material in the respective directions. The adjusted criterion is able to predict the failure of unidirectional composite in case of the mentioned combination of loading. The validation of the adjusted criterion is carried out by means of the comparison of experimental results and numerical analysis performed in finite element system MSC.Marc.

Keywords: composite, criterion, LaRC04, combination of tension and compression

1. Introduction

One way how to joint composite and metal parts is using pins. The wrapped pin joints were used in case of joint of CompoTech composite hydraulic cylinder (see Fig. 1) and metal cylinder flange, where the joint is exposed to tension. The main principle of producing the wrapped pin joints is that fibres of the composite are wrapped directly around the shape of a metal pin. It allows creating the joint without cutting fibres. Therefore, it allows creating joints that excel at strength. The tensile loading of the joint results in press in the inner diameter of the pin holes (loops), where the state of stress corresponds to loading of pressure vessel. Compression in the transverse direction (perpendicular to composite fibres) and tension in the longitudinal direction (parallel to composite fibres) reach very large values compared with ultimate strengths of material in the respective directions (see Fig. 1c).

During the investigation of the problem of pin joints in composites, several works about the joint failure were studied [1, 2, 3, 6, 11, 13]. However, only limited works were focused on the wrapped pin joints [4, 7], where the mentioned specific state of stress occurs. In work [4], the strength criterion Puck for prediction of failure of the wrapped pin joints was used. Using this criterion, major disagreement between numerical and experimental results was achieved. Also in work [7], no standard criterion for correct prediction of failure of the joint was found. Therefore, the criterion LaRC04 [12] was adjusted [7, 8].

The aim of this work is the validation of the adjusted strength criterion LaRC04 for unidirectional composite under combination of tension and compression for a wider range of geometries

---

*Corresponding author. e-mail: krystek@kme.zcu.cz.

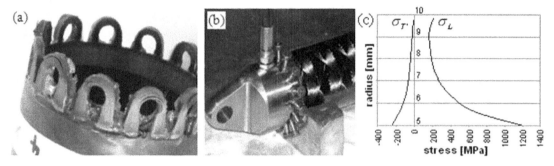

Fig. 1. (a) Composite parts of wrapped pin joint integrated in CompoTech hydraulic cylinder [7]; (b) CompoTech hydraulic cylinder [7]; (c) Stress dependences in loop [7]

than in [7, 8]. The validation of the adjusted strength criterion was carried out by means of comparison of experimental results and numerical analysis. The numerical analysis was performed in finite elements system MSC.Marc.

2. Failure criterion LaRC04

The criterion LaRC04 is a strength criterion for unidirectional composite materials developed in NASA Langley Research Center in 2004 [12]. It belongs to the group of so-called interactive criteria and so-called *direct mode* criteria. Interactive criteria include relationship between components of normal stresses and between components of normal stresses and components of shear stresses. Criterion LaRC04 is derived for full three-dimensional state of stress. Direct mode criteria distinguish several types (modes) of failure and describe each of them with independent condition. Result from criterion LaRC04 is the value of failure index. (The failure index takes the value from 0 to 1. If the failure index is equal to 1, the failure occurs.)

In the mentioned application of joint elements of the hydraulic cylinder, the three-dimensional model showed major disagreement between numerical and experimental results for failure modes LaRC04#2 and LaRC04#3 [7]. Therefore, the criterion LaRC04 was modified [7, 8]. In case of mode LaRC04#2 (matrix failure in compression in the transverse direction), it was found that the influence of stress in the fiber direction σ_L should be included in this mode. The equation for failure index of failure mode LaRC04#2 was adjusted to

$$FI_M = \left(\frac{\tau^T}{S^T - \eta^T \sigma_n + \sigma_L P_M} \right)^2 + \left(\frac{\tau^L}{S^L - \eta^L \sigma_n + \sigma_L P_M} \right)^2 \leq 1, \qquad (1)$$

where the term $\sigma_L P_M$ is the suggested adjustment. In case of mode LaRC04#3 (fiber failure in tension in the longitudinal direction), the equation for failure index was adjusted to

$$FI_F = \frac{\sigma_L}{X^T P_F} \leq 1, \qquad (2)$$

where P_F is again the suggested adjustment.

3. Sample manufacture and experiments

The special samples for the validation of adjusted strength criterion were proposed. Circular part of sample in which is achieved the investigated combination of stress will be named as loop. Design of sample and designation of geometric parameters are illustrated in Fig. 2, where d is the diameter of the pins, t is the thickness of the loop, a is the width of the loop and l is the distance between the axes of pins.

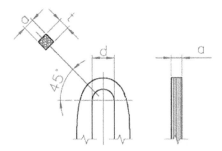

Fig. 2. Sample geometric properties [7]

The samples for experiments were produced in laboratory at Department of Mechanics, University of West Bohemia in Pilsen. Technology of filament winding was used for production of samples (Fig. 3). The fibres were impregnated and wound on the structure with pins. The samples were then hardened in oven.

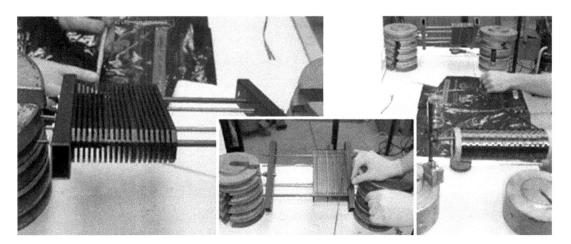

Fig. 3. Manufacture of samples

The carbon fiber T700 and epoxy resin (the resin LG120 and the hardener EM100 in ratio $100 : 34$) was used for production of the uni-directional composite samples. The fiber volume of loops was $V_f = 0.65$. Elasticity parameters in the material directions are presented in Tab. 1 and Tab. 2. Strength parameters are presented in Tab. 2.

Table 1. Elasticity parameters of composite

V_f [−]	E_L [MPa]	E_T [MPa]	$E_{T'}$ [MPa]	ν_{LT} [−]	$\nu_{TT'}$ [−]	$\nu_{T'L}$ [−]	G_{LT} [MPa]	$G_{TT'}$ [MPa]	$G_{T'L}$ [MPa]
0.65	153 730	5 940	5 940	0.335	0.332	0.013	2 755	2 230	2 755

Table 2. Strength parameters of composite

X^T [MPa]	Y^T [MPa]	Y^C [MPa]	S^L [MPa]	α_0 [°]
3 264	42	92	48	55

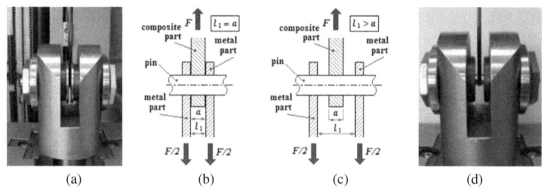

Fig. 4. (a), (b) Tight fastening; (c), (d) Free fastening

The loading was carried out by means of the machine Zwick/Roell Z050. The experimental samples were tested in tension. Each sample was fixed in the machine by means of the special components (see Fig. 4a and 4d) that were fixed in the jaws of the machine. These components allow the loading of samples with wide range of geometric parameters. Using these components, it is possible to select between two kinds of fastening of samples — tight fastening (further TF) and free fastening (further FF) (see Fig. 4). The faces of the FF loop are free. In case of the TF loop, the faces are tight fastened near the area of pin. This corresponds to fixture washers and nuts in real hydraulic cylinder.

The value of maximal tensile force was the parameter for comparison between numerical results and experimental results. The failure of fibres occurs at this maximal tensile force. The limiting factors for experiments were maximal tensile force 50 kN (maximal loading force in the used machine) and strength of metal pin with smaller diameter.

More than 150 samples with different geometries (see Fig. 5.) ($d = 8$ and 12 mm; $a = 3$ and 6 mm, $t = 1 \div 8$ mm) were tested at the temperature of $+20\,^{\circ}$C. The loading speed was $v = 0.5$ mm/min.

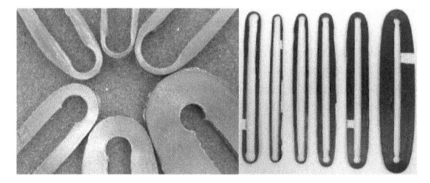

Fig. 5. Samples with different geometries

4. Failure of the samples

The matrix failure occurred prior to the fiber failure for some samples. The matrix failure divides the loop cross-section of the sample. A shape of separated part in case of the FF samples (the displacement in the pin axial direction is allowed) is obvious from Fig. 6a,b. In the separated part of cross-section, the decrease of stress in longitudinal direction σ_L occurred. Whilst, increase of stress in longitudinal direction σ_L in not separated part of cross-section occurred.

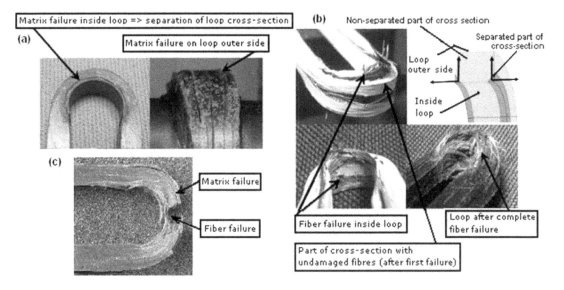

Fig. 6. (a), (b) Failure procedure of FF sample [7]; (c) Failure procedure of TF samples

Therefore, failure of fiber due to influence of tension began earlier than in case without separation of part of cross-section. The loop strength was considered as the maximal tensile force for failure of fiber (failure index of fiber $FI_F = 1$) [7, 8, 10]. "Layer separation" of the loop cross-section in case of TF samples is obvious from Fig. 6c. This separation had not the influence on the loop strength.

5. Numerical simulation

A three-dimensional numerical model was created in the finite elements system MSC.Marc. The adjusted strength criterion LaRC04 was implemented in this finite element system [9]. The model was created from hexahedral elements with 8 nodes (SOLID elements). Regarding symmetry of the sample, only one eighth of the sample was modelled (see Fig. 7b). Orthotropic material properties were assigned to the elements respecting the orientation of the fibers. The loading was controlled by the displacement of a rigid surface, which simulates the pin (see Fig. 7a.).

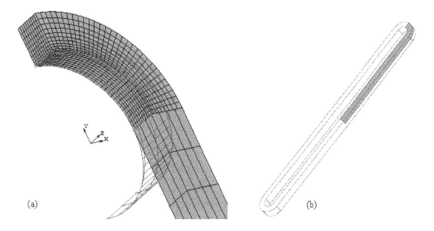

Fig. 7. (a) Mesh of numerical model; (b) Modelled one eight of sample

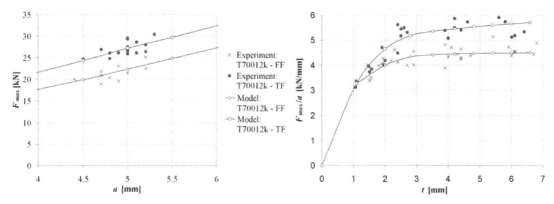

Fig. 8. Comparison of experimental and numerical results for $d = 10$ mm, $a = 5$ mm, $V_f = 0.50$ [7]

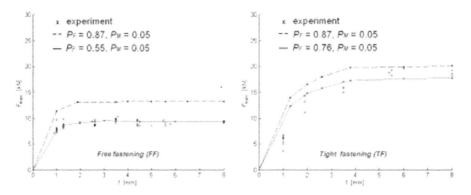

Fig. 9. Comparison of experimental and numerical results for $d = 8$ mm, $a = 3$ mm, $V_f = 0.65$

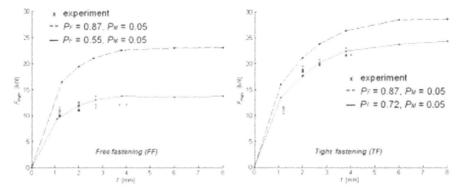

Fig. 10. Comparison of experimental and numerical results for $d = 12$ mm, $a = 3$ mm, $V_f = 0.65$

The whole procedure of numerical simulation was automated. The parametrical creating of the numerical model of sample was created by means of scripts. These scripts were generated in system Matlab from designed geometric and material parameters.

The results of finite elements analysis (FEA) (with the adjusted criterion LaRC04) and tensile tests are shown in Fig. 8 [7]. This analysis was performed for samples with diameter $d = 10$ mm and fiber volume of loop $V_f = 0.50$. The best agreement between experimental and numerical results was achieved for investigated geometries ($d = 10$ mm, $t = 1 \div 6.8$ mm, $a = 4.4 \div 5.8$ mm) of samples and the fiber volume of loop $V_f = 0.50$ for parameters (non-dimensional) $P_M = 0.05$ and $P_F = 0.87$ [7].

The comparison of the experimental and numerical results is shown in Fig. 9 and Fig. 10. The graphs show the dependences of the sample strength (maximal tensile force $F_{\max}$) on the sample thickness t for diameter $d = 8$ mm (see Fig. 9) and diameter $d = 12$ mm (see Fig. 10). The dashed curves are displayed for original parameters ($P_M = 0.05$ and $P_F = 0.87$). The solid curves are displayed for the new parameters. Using the new parameters, the best agreement between numerical and experimental results was achieved. The new values of the parameters are presented in the graphs. The fiber volume of loops was $V_f = 0.65$ and width of loop was $a = 3$ mm.

From mentioned results, good agreement between experimental results and numerical analysis was verified for adjustment of failure mode LaRC04#2 (matrix failure in compression in the transverse direction). This adjustment involves influence of stress in the longitudinal direction. The modification of failure mode LaRC04#2 for parameter $P_M = 0.05$ is correct also in the case of pure compression in transverse direction or when stress in fiber direction is low, because the influence of the adjustment is very low in this case.

In case of failure mode LaRC04#3, the parameter P_F is different for different fiber volumes, i.e. different material properties and different geometry of sample. It is necessary to determinate these parameters for specific material and geometry. Technique of identification of these parameters will be investigated in the future work.

The dependences of maximal tensile force (strength of sample) on combinations of two various geometrical parameters (third is constant) in case of TF samples from finite element analysis are shown in Fig. 11. Significant influence of width of sample a is obvious from those dependences. The sample strength increases proportionally with the width a. The influence of the loop thickness t is significant only up to determinate thickness (e.g. thickness $t = 4$ mm for diameter $d = 8$ mm). The influence of the sample thickness t on its strength increases with increasing diameter d (see Fig. 11b). In case of TF loops, the gradual linearization of the dependence of sample strength on inner diameter d occurs with increasing loop thickness (see Fig. 11b).

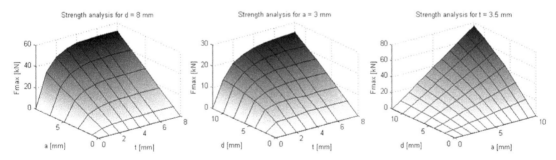

Fig. 11. FEA of TF samples for (a) $d = 8$ mm, (b) $a = 3$ mm, (c) $t = 3.5$ mm

6. Conclusion

In the presented work, validation of the adjusted strength criterion LaRC04 for unidirectional composite under combination of tension and compression by means of comparison of experiments and numerical models for a wide range of samples was carried out. The numerical strength analysis was performed in finite element system MSC.Marc. The experimental samples were tested in tension. For the experiments, more than 150 samples with different geometries were manufactured and tested.

The results show good agreement between experimental results and numerical analysis for adjustment of failure mode LaRC04#2 (matrix failure in compression in the transverse direction), this adjustment involves influence of stress in the longitudinal direction, for the investigated range of samples.

In case of failure mode LaRC04#3, the parameter P_F is different for different fiber volume and different geometry of sample. It is necessary to determinate these parameters for specific material and geometry. Technique of identification of these parameters will be investigated in the future work.

Acknowledgements

This paper includes partial results from the research project MŠMT 1M0519 — Research Centre of Rail Vehicles supported by the Czech Ministry of Education and partial results from the research project GAAS CR no. A200760611.

References

[1] Atas, C., Bearing strength of pinned joints in woven fabric composite with small weaving angles. Composite Structures 2009, 88, 40–45.

[2] Dano, M. L., Gerdron, G., Picard, A., Stress and failure analysis of mechanically fastened joints in composite laminates. Composite Structures 2000, 85, 287–296.

[3] Dano, M. L., Kamal, E., Gendron, G., Analysis of bolted joins in composite laminates: Strains and bearing stiffness predictions. Composite Structures 2007, 79, 562-570.

[4] Havar, T., Beitrag zur Gestaltung und Auslegung von 3D-Verstärkten Faserverbundschlaufen. Institut für Flugzeugbau der Universität Stuttgart, Stuttgart, 2007.

[5] Hinton, M. J., Kaddour, A. S., Soden, P. D., Failure criteria in fibre reinforced polymer composites: The World-Wide Failure Exercise. Elsevier, 2004.

[6] Kishore, A. N., Malhotra, S. K., Prasad, N. S., Failure analysis of multi-pin joints in glass fibre/epoxy composite laminates. Composite Structures 2009, 91, 266–277.

[7] Kottner, R., Joining of composite and steel machine parts with special focus on stiffness and strength (in Czech). Doctoral thesis, University of West Bohemia, Plzeň, 2007.

[8] Kottner, R., Kroupa, T., Krystek, J., Laš, V., Strength prediction of composite part of wrapped pin joint of composite/metal using FEA. Mechanical response of composite, Imperial College London, UK, 2009.

[9] Kroupa, T., Damage of composites due to impact (in Czech). Doctoral thesis, University of West Bohemia, Plzeň, 2007.

[10] Krystek, J., Kottner, R., Kroupa, T., Laš, V., Influence of geometric parameters of composite part of composite/metal wrapped pin joint on its strength. Machine Modelling and Simulation, Scientific and Technical Society at the University of Žilina, Slovensko, 2009.

[11] Okutan, B., The effects of geometric parameters on the failure strength for pin-loaded multidirectional fiber-glass reinforced epoxy laminate. Composites: Part B 2002, 33, 567–578.

[12] Pinho, S. T., Dávila, C. G., Camanho, P. P., Iannucci, L., Robinson, P., Failure Models and Criteria for FRP Under In-Plane or Three-Dimensional Stress States Including Shear Non-Linearity. Research report, NASA/TM-2005-213530, NASA Langley Research Center, 2005.

[13] Withworth, H. A., Othieno, M., Barton, O., Failure analysis of composite pin loaded joints. Composite Structures 2003, 59, 261–266.

Permissions

All chapters in this book were first published in ACM, by University of West Bohemia; hereby published with permission under the Creative Commons Attribution License or equivalent. Every chapter published in this book has been scrutinized by our experts. Their significance has been extensively debated. The topics covered herein carry significant findings which will fuel the growth of the discipline. They may even be implemented as practical applications or may be referred to as a beginning point for another development.

The contributors of this book come from diverse backgrounds, making this book a truly international effort. This book will bring forth new frontiers with its revolutionizing research information and detailed analysis of the nascent developments around the world.

We would like to thank all the contributing authors for lending their expertise to make the book truly unique. They have played a crucial role in the development of this book. Without their invaluable contributions this book wouldn't have been possible. They have made vital efforts to compile up to date information on the varied aspects of this subject to make this book a valuable addition to the collection of many professionals and students.

This book was conceptualized with the vision of imparting up-to-date information and advanced data in this field. To ensure the same, a matchless editorial board was set up. Every individual on the board went through rigorous rounds of assessment to prove their worth. After which they invested a large part of their time researching and compiling the most relevant data for our readers.

The editorial board has been involved in producing this book since its inception. They have spent rigorous hours researching and exploring the diverse topics which have resulted in the successful publishing of this book. They have passed on their knowledge of decades through this book. To expedite this challenging task, the publisher supported the team at every step. A small team of assistant editors was also appointed to further simplify the editing procedure and attain best results for the readers.

Apart from the editorial board, the designing team has also invested a significant amount of their time in understanding the subject and creating the most relevant covers. They scrutinized every image to scout for the most suitable representation of the subject and create an appropriate cover for the book.

The publishing team has been an ardent support to the editorial, designing and production team. Their endless efforts to recruit the best for this project, has resulted in the accomplishment of this book. They are a veteran in the field of academics and their pool of knowledge is as vast as their experience in printing. Their expertise and guidance has proved useful at every step. Their uncompromising quality standards have made this book an exceptional effort. Their encouragement from time to time has been an inspiration for everyone.

The publisher and the editorial board hope that this book will prove to be a valuable piece of knowledge for researchers, students, practitioners and scholars across the globe.

List of Contributors

M. Mikhov
Faculty of Automatics, Technical University of Sofia, 8 KlimentOhridski Blvd., 1757 Sofia, Bulgaria

M. Zajíček, V. Adámek and J. Dupal
Faculty of Applied Sciences, University of West Bohemia, Univerzitní 22, 306 14 Plzeň, Czech Republic

M. Ševčík and L. Náhlík
Institute of Physics of Materials, Czech Academy of Sciences, Žižkova 22, 616 62 Brno, Czech Republic
Institute of Solid Mechanics, Mechatronics and Biomechanics, Brno University of Technology, Technická 2, 616 69 Brno, Czech Republic

P. Hutař and Z. Knésl
Institute of Physics of Materials, Czech Academy of Sciences, Žižkova 22, 616 62 Brno, Czech Republic

J. Očenášek and J. Voldřich
New Technologies Research Centre, University of West Bohemia, Univerzitní 8, 306 14 Plzeň, Czech Republic

V. Kleisner and R. Zemčík
Faculty of Applied Sciences, University of West Bohemia, Univerzitní 22, 306 14 Plzeň, Czech Republic

S. Seitl and Z. Knésl
Institute of Physics of Materials, Academy of Sciences of the Czech Republic, v.v.i.Žižkova 22, 616 62 Brno, Czech Republic

V. Veselý and L. Řoutil
Institute of Structural Mechanics, Faculty of Civil Engineering, Brno University of Technology, Veveří 331/95, 602 00 Brno, Czech Republic

P. Polach and M. Hajžman
Section of Materials and Mechanical Engineering Research, Výzkumný a zkušebníústavPlzeň s. r. o., Tylova 1581/46, 301 00 Plzeň, Czech Republic

Z. Šika
Department of Mechanics, Biomechanics and Mechatronics, Faculty of Mechanical Engineering, Czech Technical University in Prague, Technická 4, 166 07 Praha, Czech Republic

M. Byrtus, M. Hajžman and V. Zeman
Faculty of Applied Sciences, University of West Bohemia, Univerzitní 22, 306 14 Plzeň, Czech Republic

J. Klusák
Institute of Physics of Materials, Academy of Sciences of the Czech Republic, Žižkova 22, 616 62 Brno, Czech Republic

T. Profant and M. Kotoul
Brno University of Technology, Technická 2, 616 69 Brno, Czech Republic

V. Veselý and P. Frantík
Faculty of Civil Engineering, BUT Brno, Veveří 331/95, 602 00 Brno, Czech Republic

J. Svoboda, M. Balda and V. Fröhlich
Institute of Thermomechanics, Czech Academy of Sciences, Veleslavínova 11, 301 14 Plzeň, Czech Republic

R. Melicher
Faculty of Applied Mechanics, Žilina university in Žilina, Univerzitná 1, 010 26 Žilina, Slovak Republic

P. Polach and M. Hajžmana
Section of Materials and Mechanical Engineering Research, ŠKODA VÝZKUM, s. r. o., Tylova 1/57, 316 00 Plzeň, Czech Republic

J. Soukup and J. Volek
Department of Mechanics and Machines, Faculty of Production Technology and Management, University of J. E. Purkyně in Ústínad Labem, Na Okraji 1001, 400 96 Ústínad Labem, Czech Republic

V. Zeman and Z. Hlaváč
Faculty of Applied Sciences, University of West Bohemia, Univerzitní 22, 306 14 Plzeň, Czech Republic

H. Kocková
New Technologies – Research Centre in the Westbohemian Region, University of West Bohemia, Univerzitní 22, 306 14 Plzeň, Czech Republic

R. Cimrman
Department of Mechanics, Faculty of Applied Sciences, University of West Bohemia, Univerzitní 22, 306 14 Plzeň, Czech Republic

J. Horáček
Institute of Thermomechanics, Czech Academy of Sciences, Dolejškova 5, 182 00 Praha 8, Czech Republic

P. Sváček
Institute of Thermomechanics, Czech Academy of Sciences, Dolejškova 5, 182 00 Praha 8, Czech Republic
Dep. of Technical Mathematics, Faculty of Mechanical Engineering, Czech Technical University in Prague, Karlovonam. 13, 121 35 Praha 2, Czech Republic

J. Krystek, R. Kottner, L. Bek and V. Laš
Faculty of Applied Sciences, University of West Bohemia, Univerzitn´ı 22, 306 14 Plzeň, Czech Republic

Index

A

Absolute Nodal Coordinate Formulation (ancf), 71

Accumulative Roll Bonding (arb), 132

Activation Control, 167, 169, 171, 173, 175, 177, 179

Aluminium Workpiece, 131, 133, 137, 139-141

Analytical Solution, 9-10, 16-19, 32, 134, 143-144, 153

Arbitrary Lagrangian-eulerian Method, 181

Artificial Test Track, 143

B

Bi-material Notches, 99, 106

Biogenous Filter Cake, 31-33, 35, 37, 39, 41

Biomechanics, 21, 52, 69, 80, 154, 167, 181, 189, 191

Bogie Frame (bf), 155

Bumper Reinforcement, 43-45, 47, 49-52

C

Calcium Activation, 167, 171, 179

Cohesive Crack Model, 107

Combined Loading, 121, 127, 129

Composite Materials, 43, 45-47, 52, 193-194

Contraction Cycle, 168

Crack Propagation, 21-22, 24, 30, 58, 106, 110, 116, 119, 122, 124-127, 129

Crack Tip Stress Field, 53, 55, 57, 59, 61, 63, 65-67

Cross-bridge Distribution, 167, 171, 173, 179

Crucial Parameters, 69, 71, 73, 75, 77, 79

D

Discretization Methodology, 21

Drive Systems, 1, 3, 5, 7-8

Dynamic Simulation, 1, 3, 5, 7-8

E

Eddy-currents, 2

Elastic Matrix, 45

Electromechanical Systems, 1

Equal Channel Angular Pressing (ecap), 131-132

Euler-bernoulli Beam Theory, 9

F

Fiber Architecture, 167, 179

Finite Element (fem), 131

Finite Element Method, 20, 22, 43, 53, 59, 80, 142, 181, 184, 189, 191

Fractionation Mechanism, 33, 35

Fracture Mechanics, 21-23, 29-30, 53-55, 59, 65-68, 99, 104, 106-108, 111, 118-121

Fracture Parameters, 26, 53-55, 59, 63, 66, 68, 99, 118

Fracture Process Zone, 55, 68, 107, 110-113, 115, 117-120

Fracture Process Zone (fpz), 55, 107

Functionally Graded Materials (fgm), 21

G

Generalized Stress Intensity Factor (gsif), 99

H

High Pressure Torsion (hpt), 132

Homogenous Materials, 21, 23

Hydraulic Cylinder, 193-194, 196

I

Inverted Pendulum, 69-73, 75-77, 79-80

K

Kelvin-voight Material, 9

Kinematic Excitation, 69, 71-74, 76, 79, 143-145, 151

L

Linear Compression Chamber, 31, 33, 36, 39, 41

Linear Elastic Fracture Mechanics, 21-23, 55, 67, 107-108, 111

M

Mathematical Modeling, 1-5, 7, 31, 33, 35, 37, 39, 41, 179

Multibody Model, 143-144, 149-151, 153-154

N

Nanostructured Materials (nsm), 131

Navier-stokes Equations, 181, 183, 191

Non-stationary Problems, 9, 11, 17, 19

Nonlinear System, 69, 71, 86

Numerical Approximation, 181-183, 185, 187, 189, 191

Numerical Estimation, 21, 99, 101, 103, 105

Numerical Investigation, 143, 145, 147, 149, 151, 153-154

Numerical Simulation, 31, 39, 43, 45-49, 53, 59, 81, 115-116, 131, 133, 137, 139, 141-143, 151, 180-182, 191-192, 197-198

O

Oilseed Material, 31, 33, 35, 37, 39, 41

P

Passive Safety, 43, 52

Permanent Magnet Synchronous Motors, 1, 8

Plastic Deformation, 121, 131, 133, 137, 139-142

Power Spectral Density (psd), 121

Power-law Material Change, 21

Q

Quasi-brittle Materials, 53, 55, 66, 68, 107-108, 110-113, 115, 117-120

R

Railway Vehicle, 155, 157, 159, 161, 163-166

Rankine Strength Criterion, 107, 118

Rcar (research Council for Automobile Repairs), 43

Reconstruction, 107-108, 111-113, 115, 117, 119-120

Rheological Properties, 31, 41

S

Severe Plastic Deformation (spd), 131

Short-circuit Motor Moment, 155, 157, 159, 161, 163, 165

Skeletal Muscle Model, 167, 169, 171, 173, 175, 177, 179

Spectral Properties, 121, 123, 125, 127, 129-130

Stator Currents, 2

Strain Energy Density Factor (sedf), 101, 105

Strength Criterion, 107, 118, 193-195, 197, 199

Stress Intensity Factor, 21-23, 25-29, 54, 58-64, 67, 99, 104, 125-126, 129

Symmetric Channel, 181, 183, 185, 187, 189, 191

T

Tangential Stress, 105

Tensile Failure, 107-108, 111, 113, 115, 117, 119

Timoshenko Theory, 9-10, 12, 17, 19

Tip Stress Field, 22, 53, 55-57, 59, 61, 63, 65-67

Torsion Loading, 121-123, 125, 127, 129

Traction Motor (tm), 155

Trolleybus, 143-147, 149-154

V

Vertical Dynamics, 143, 145, 147, 149, 151, 153-154

Vibrating Walls, 181-183, 185, 187, 189, 191

Viscoelastic Orthotropic Beams, 9, 11, 17, 19

W

Wedge-splitting Specimens, 53

Wheelset Drive Load, 155, 157, 159, 161, 163, 165